Penicillin
Meeting the Challenge

Prof. Sir Alexander Fleming (1881–1955), M.B., B.S., F.R.C.P., F.R.C.S., F.R.S.

Penicillin

Meeting the Challenge

Gladys L. Hobby

Yale University Press
New Haven and London

Designed by James J. Johnson and set in Times Roman
type by Graphicraft Typesetters Ltd. Hong Kong.
Printed in the United States of America by
Vail-Ballou Press, Binghamton, N.Y.

Library of Congress Cataloging in Publication Data

Hobby, Gladys L., 1910–
    Penicillin: Meeting the Challenge.

    Bibliography: p.
    Includes index.
    1. Penicillin—History. I. Title.
RS165.P38H63 1985   615'.32923   84–19689
ISBN 0–300–03225–0 (alk. paper)

*The paper in this book meets the guidelines for
permanence and durability of the Committee on
Production Guidelines for Book Longevity of the
Council on Library Resources.*

10 9 8 7 6 5 4 3 2 1

To W. H. E.

Contents

Contents

List of Illustrations

List of Figures and Tables

Foreword

In 1939, ten years after Alexander Fleming had shared his observations on penicillin with the scientific public, this antimicrobial agent was little more than another laboratory curiosity. Catalyzed by the results of the experimental and clinical studies of Florey and his coworkers, interest in penicillin as a therapeutic agent took off in 1940 at a meteoric pace. A little more than five years later, by 1945, the far-reaching potential of penicillin for successful treatment of life-threatening diseases (and some that were less than life threatening, but nonetheless important) was firmly established. With few competitors at the time, penicillin was heralded as the most potent of all agents for treatment of acute infectious diseases. Now, almost forty years later and with many competitors, it is still so regarded. Incidentally, many of these competitors, those in the penicillin-cephalosporin categories, owe their existence to application of the methodologies of penicillin production developed in the 1940–45 era.

The history of penicillin has been the subject of at least thirteen published books. In the majority of these, the parts played by Fleming, Florey, and their coworkers in the development of this antimicrobial have been the primary concern. In the most recent books, the focus has been on the elucidation of the structure of penicillin and its total synthesis. None of these histories has dealt adequately with the development of penicillin as a therapeutic agent, how this development was effected, and what it has meant to the continued search for more effective antimicrobial drugs.

Dr. Hobby's book is concerned with these subjects. It delineates the diverse forces that contributed to the long gap between Fleming's pioneering observations and critical evaluation of the therapeutic potentials of penicillin. It covers the modest beginnings of such evaluations on both sides of the Atlantic and how they were serviced by the halting efforts of individual groups of investigators to produce penicillin. It shows why, in the stress of the bombings of Britain during World War II, the British pharmaceutical and chemical companies were not able to meet the challenge for supplies of penicillin generated by the therapeutic accomplishments of Florey. It tells of the latter's turn to the United States for assistance not available at home. It describes the responses of governmental organizations, pharmaceutical companies, research institutes, university laboratories, and clinics in the United States to this request, responses that took the form of a collaborative effort of scope and intensity not attempted either before or since in any area of biological science. It shows how these responses made it possible to identify in something less than three years almost all the major uses of penicillin, uses still recognized forty years later, and to provide penicillin in quantities appropriate to all foreseeable needs at a price that even the poor could pay. Fortunately, the account does not stop there, but goes on to show how various facets of the fermentation technology essential for production of penicillin have underpinned development of the many penicillins and cephalosporins that are ours to use today, setting the stage for the future.

Such coverage as Dr. Hobby has attempted would never have been possible without personal involvement in work on penicillin from its humble beginnings in the laboratory, through successive clinical trials, through the steps in production that made these trials pursuable on a meaningful scale, and on to the continuing search for new and more effective antimicrobials. Her personal involvement as a scientist in this work has given her insights and perspectives that no other individual involved with the development of penicillin and alive today can match. Her commitments as a scientist to accuracy and precision have compelled her to document essentially every important point made or issue raised. This documentation adds much to the lasting value of this book.

Whereas documentation sometimes destroys the spontaneity of a written work, this has not happened in this book. It is filled with exciting events. One in particular stands out. It was the social gathering at John Fulton's house in New Haven during which there was a brief interchange between Howard Florey and Ross Harrison that set the stage for contacts with the U.S. Department of Agriculture Regional Research Laboratory in

Peoria, Illinois. It was this casual contact that made quantity production of penicillin possible in such a remarkably short time.

Readers of this book will doubtless agree that it is aptly titled. What better than *Penicillin: Meeting the Challenge* to encompass those phases of the work that made penicillin available for the studies that delineated its therapeutic potential and available for later widespread use—except perhaps "They Dared to Venture" and that most descriptive addition, *"and they succeeded."*

LEON H. SCHMIDT

Preface

The discovery of penicillin ranks among the most significant discoveries of mankind. The story of this discovery has been told many times in "as I remember it" tales, with their individual viewpoints; but no account has been complete, none well documented, and none adequately discusses the contributions made by numerous people in the United States and the United Kingdom, without whom penicillin never would have become a usable and effective drug.

This book is not the story of the discovery of penicillin, nor is it a biography of Alexander Fleming, its sole discoverer. Neither is it a biography of Howard Florey nor an account of his contributions, or those of his Oxford associates, to penicillin's development. The lives and contributions of these men have been recorded previously.

This book is the story of penicillin itself and the hundreds of people who worked together to isolate and characterize the substance and to raise it from a laboratory curiosity to the first of a remarkable series of highly effective chemotherapeutic drugs. It is the story of one of the greatest ventures in group research that has ever been conducted—the story of an unusually successful collaborative venture.

To understand what transpired between 1928 when penicillin was discovered and 1946 when it became available for widespread clinical use, one must understand the scientific climate of that period. The immunological successes of the late nineteenth and early twentieth centuries and the success achieved with vaccine therapy had conditioned the scientific community against ready acceptance of agents such as penicillin. Until the late 1930s, few believed that infectious diseases of bacterial origin could be effectively treated by chemical agents, and fewer still believed that effective chemotherapeutic agents could be derived from the growth products of other microorganisms. Even the apparent effectiveness of salvarsan (Ehrlich's 606, the "magic bullet" for treatment of syphilis) was not sufficient to convince the scientific community. Not until 1935, when Gerhard Domagk, director of the Institute of Experimental Pathology at Elberfeld, a part of the large research complex of I. G. Farbenindustrie in Germany, published the results of studies carried out three years earlier, showing the remarkable therapeutic effectiveness of prontosil against hemolytic streptococcal infections, was the tide of opinion turned.

More than fifty years have elapsed since Alexander Fleming's discovery of penicillin, which preceded Domagk's observations on prontosil by several years. Fleming recognized that the lytic action of penicillin was of biological importance, but limited knowledge of microbial physiology and biochemical mechanisms and inadequate analytical techniques, among other things, precluded production of sufficient quantities of the active substance for studies needed to prove its potential. Only now, after half a century, can one view with perspective what transpired in those early years and fully recognize the significance of the event.

The story told here is based on a review and analysis of both published and unpublished scientific data and reports. Documentation destroys the flavor of the story. It eliminates the spontaneity, the excitement, the joy, and the amazement that accompanied each new development. It even leads one to forget that penicillin was a British discovery—that it was discovered in England, first studied in England, first used clinically in England—and that it probably was one of Britain's major wartime contributions to society, second only to its provision of an operational base from which Hitler and the Third Reich were destroyed. It is easy to forget all this, for the studies came to fruition largely through the contributions of investigators in the United States.

But documentation is long overdue. Too many myths and unfounded tales have been passed along, from person to person, during penicillin's fifty-year history. It is my hope that this book will provide a valid and comprehensive record of one of the most important scientific developments in history.

Acknowledgments

When I started research on this book, I contacted more than one hundred persons who played an important role in penicillin's early development. Eventually the number was multiplied many times. Almost everyone at first assured me that records had been discarded long ago, that they could remember nothing. But as time passed, memories were stimulated, and it became clear that—by design or by accident, perhaps because of a bit of sentiment or personal pride—most had retained a major portion of their records. And none had forgotten what had transpired during that important and exciting period.

Copies of correspondence, reports, and official and informal documents started coming to me, often from very remote sources. There was much duplication, and I was (and am) constantly amazed at the wide circulation given to what was then considered the most confidential information. Most interesting were the personal recollections of those who were among the first to work on penicillin, or to be treated with it, and those who had personal encounters with Alexander Fleming, Howard and Ethel Florey, Ernst and Anne Chain, Henry Dawson, Karl Meyer, Norman Heatley, and the many others whose names quickly come to mind as one thinks back on the early days of penicillin. Even today, after five years of research for this book, scarcely a week passes that some new piece of information concerning the "wonder drug" and its early development does not come to me. Unfortunately, I cannot include everything in the present volume. I am grateful, nonetheless, to each person who took time to send material to me.

I owe special thanks to my longtime friend, the late Dr. Walsh McDermott of the Cornell University Medical College and the Robert Wood Johnson Foundation, who reviewed with me many of the records received and devoted many hours to reflecting with me on what has transpired since 1940 and on what should be included in this historical review. I am grateful also to Dr. Leon Schmidt of the University of Alabama School of Medicine who shared with me his knowledge of chemotherapy acquired during more than forty years of research on chemotherapeutic drugs and carefully reviewed several drafts of the manuscript. Likewise I am indebted to Dr. Ralph Tompsett, emeritus professor of medicine and chief of infectious diseases at Baylor University Hospital and Medical School, and to Dr. Robert Coghill, formerly chief of the Fermentation Division of the Northern Regional Research Laboratory in Peoria, Illinois, who also reviewed the manuscript.

I owe much to my many friends at Pfizer, Inc. who encouraged my undertaking the project, gave me free access to their files, and assisted in more ways than one can mention. Mr. E. T. Littlejohn and Mr. Charles Fry, formerly of Pfizer's Division of Public Affairs, helped in initial stages of the project. Subsequently, Mr. Robert A. Wilson and Dr. Donald Jacob of the same division contributed much. Dr. Robert Feeney, currently vice-president of licensing and development, the late Dr. Frank Murphy of the Legal Division, Jasper Kane and Alexander Finlay, both active from the start in biochemical research on penicillin and its development, and Mr. George B. Stone who from the beginning witnessed each stage of penicillin's development and recently prepared a four-volume compilation of the company's most important records—all these people, among many others, provided important information. The unlimited support of Mr. E. T. Pratt, Jr. (chairman of the board) and Dr. Gerald Laubach (president) in particular has been most valuable.

Much credit is due Mr. Fred Stock, formerly chief, Drugs and Cosmetics Branch of the War Production Board; the late Sir Ernst Chain, Dr. Norman Heatley, and Sir Edward Abraham at the Sir William Dunn School of Pathology in Oxford, England; Lady Margaret Florey, who gave me access to the Florey Archives at the Royal Society; the late Dr. Lawrence P. Garrod, Mrs. John Fulton, Lady Anne Chain; Dr. Boyd Woodruff of Merck & Co., Inc.; Dr. John T. Sheehan of E. R. Squibb & Sons, Inc.; and, among many others, Dr. Peter Regna who reviewed for me the sections pertaining to the chemical structure and synthesis of penicillin. Credit is due also to Miss Louise Good, Miss Anne Blevin, and

Dr. Erna Alture-Werber who loaned laboratory records on patients with subacute bacterial endocarditis treated more than forty years ago by the late Drs. M. H. Dawson and Ward MacNeal of the Presbyterian and Postgraduate Hospitals in New York City and by the late Dr. Leo Loewe of the Jewish Hospital in Brooklyn.

Special thanks are due Sir Austin Bide, chairman, and Mr. Alan Raper, director of Glaxo Holdings, Ltd., and to Mr. Bryan Emery, who at the request of Sir Austin instigated a search of Glaxo's records and compiled for me an account of what had been done many years ago at the company. Special thanks are due also to Dr. W. R. Boon, Dr. W. G. M. Jones, Dr. J. A. Robinson, Dr. R. Gow, the late Dr. A. Spinks, and Mr. J. M. Walthew of Imperial Chemical Industries who loaned copies of early ICI files for my review. In addition, I owe thanks to Dr. James Morrison and Mr. L. M. Miall of Pfizer, Ltd., the latter of whom furnished information on his personal experiences in making penicillin at the Kemball, Bishop plant in the early 1940s; to Dr. John Hastings of Distillers Corporation; to Dr. Ralph Batchelor and Dr. George Rolinson of Beecham Laboratories; and to Dr. James Cain of the National Research Development Corporation, London.

Great credit is due Miss Veronica Plucinski, librarian at Pfizer's Library for Professional Information, for finding many obscure articles and references. Without her enthusiastic cooperation, much material could not have been evaluated firsthand. Similarly, Miss Robin Green of Pfizer's Department of Public Affairs provided me with important support services.

My thanks are due also to: Mr. C. E. Dewing and Mr. Lee Johnson of the U.S. National Archives, Washington, D.C.; Mr. Ferenc Gyorgyey, librarian at the Yale Medical School Historical Library; Mrs. Patricia Stark, librarian at the Manuscript Department of the Sterling Memorial Library, Yale University; Mr. N. H. Robinson, librarian at the Royal Society; Mr. Kellihan and Mr. Bourke, manuscript librarians at the British Library at the time my study of the Fleming records began, and Mr. D. P. Waley, current keeper of manuscripts at the British Library; Miss Sophie Clark of the Medical Research Council; Mr. Gene E. McCormick, former corporate historian at Eli Lilly & Company; and Dr. Michael Gregg, deputy director in 1978 of the Bureau of Epidemiology of the U.S. Center for Disease Control who guided me to records on morbidity and mortality rates prior to the use of penicillin in the United States.

Lady Amalia Fleming has given wholehearted support to this project, has reviewed portions of the manuscript, and has approved for use all information cited from Sir Alexander Fleming's records. I am grateful to Lady Fleming for her constant interest and assistance.

I wish to thank also Mr. Theodore Hetzel of Kennett Square, Pennsylvania, who reproduced for me some of the oldest and most "brown-with-age" illustrations, and Mr. Fred Mileshko of Pfizer, Inc., who obtained for me most other illustrations.

I am pleased to mention Miss Eileen Luck who typed early portions of the manuscript and Mrs. Barbara Merkt who later typed draft after draft of the entire book. Their facility with word processing equipment has helped enormously. Mrs. Merkt, moreover, contributed much to completion of the book by her repeated cross-referencing of materials included and by preliminary editing of the manuscript.

Finally, I am most grateful to Mr. Edward Tripp, editor-in-chief of Yale University Press, without whose knowledge, expertise, and cooperation this book could not have been completed; to Mrs. Ellen Graham, who at its inception supported my undertaking to write it; to Charlotte Dihoff, who skillfully ushered the manuscript through the multiple stages of production; and to all members of the Yale University Press for their patience, interest, and assistance.

PART I

England's Great Wartime Contribution to Society

1

The Discovery, 1928

Alexander Fleming discovered penicillin in 1928.[1] He named it, described its properties, and suggested cautiously that the use of the substance as a laboratory tool might be secondary in importance to its possible use in the treatment of bacterial infections:

Penicillin, in regard to infections with sensitive microbes appears to have some advantages over the well-known chemical antiseptics.... Experiments in connection with its value in the treatment of pyogenic infections are in progress.... In addition to its possible use in the treatment of bacterial infections penicillin is certainly useful to the bacteriologist for its power in inhibiting unwanted microbes in bacterial cultures.[1]

Fleming, who was professor of bacteriology at St. Mary's Hospital Medical School in London, made his first observations on penicillin just six years after he had discovered another lytic agent to which he had given the Greek name *lysozyme*.[2] Bacteriolytic agents are substances that act on the carbohydrate moiety of certain bacteria, causing their disintegration. The discovery of lysozyme is regarded by some as more fundamentally important than the discovery of penicillin.

Background of the Discovery

The discovery of penicillin was announced in the scientific literature in June 1929, nine months after the event, but the article aroused no special interest. Indeed, the discovery, regarded merely as an observation, remained virtually unknown outside scientific circles, for the scientific community at the time considered publicity to be unethical. Not until more than a decade later was penicillin rediscovered through a survey of the literature on microbial antagonisms made by Ernst Chain.

The question is still asked: Why did it take more than ten years for someone to recognize the significance of Fleming's discovery and to begin to develop penicillin as a chemotherapeutic agent?

Actually, some believe that not ten, but more than forty years may have elapsed from the time of the first observations of penicillin until the recognition of its importance. For in 1896, a young medical student, Ernest Augustin Duchesne at the École du Service de Santé Militaire, in Lyon, France, had studied the survival and growth of bacteria and molds, separately and together, when suspended in sterile water, in contaminated water, and in a nutrient medium.[3,4] In some experiments, strains of *Escherichia coli* or typhoid bacilli were employed; in others, a strain of the mold *Penicillium glaucum* was used. Duchesne, recognizing from the beginning that there are multiple reasons for microbial antagonism, stated that some organisms may require oxygen to survive and grow, that others may have special nutritional requirements, and that still others may secrete substances toxic to other organisms.

No mention was made by Duchesne, however, of having actually demonstrated a substance with antibacterial properties. In a thesis entitled "Contribution à l'étude de la concurrence vitale chez les microorganismes: Antagonisme entre les moisissures et les microbes," he commented only that certain molds, including *Penicillium glaucum*, when inoculated into an animal at the same time as very virulent cultures of certain pathogenic organisms (*E. coli* and typhoid bacilli), were capable of attenuating to a notable degree the virulence of these bacterial cultures.[3]

Another indication that Fleming was not the first to observe the action of penicillin is contained in a letter written in 1946 by Professor D. A. Gratia of the University of Liège to Professor John Fulton at Yale University. In it, Gratia expressed surprise at reading in a French magazine the translation of an article by Fulton in which Gratia's work on antibiotics was mentioned. He went on to say:

In one of the short notes I published as far [back] as 1925, there is for instance a [description] of an observation of the bacteriolytic action of a penicillium on anthrax bacilli which is identical to the initial observation published by Fleming in 1929. I gave the observation as an example of the general character of the necessary destruction of bacteria by molds or other specialized bacteria. If I did not at the time, study immediately that particular case of antibiosis by Penicillium, it is only because I was already involved [in] other similar instances such as the antibiotic action of Actinomyces or of B. Coli. Then I very unfortunately fell sick and had to abandon work before I could start the study of Penicillium. When I resumed work, in 1929, my active strain of Penicillium was lost and...Penicillin [had] just [been] discovered by Fleming: As the latter is a friend of mine and as my observation, although [earlier], was not a contribution of enough importance to the knowledge of Penicillin itself, I never wanted to make any claim of priority in the discovery of Penicillin. But when we come to the general significance of antibiosis of which Penicillin is only a particular instance, then I could have expected to see my studies on "Streptothrix" and my observation on Penicillium recognized as the anticipation, the immediate [forerunner] of Penicillin and of streptomycin. It seems that this position is progressively coming [to] light after having remained in oblivion. I know that now Florey, Chain as well as Waksman recognise that they have been decidedly inspired by my work.[5]

In the end of the nineteenth century and the beginning of the twentieth, the new branch of science called bacteriology began to develop.[6-8] In 1881, Robert Koch published his methods for the study of pathogenic microorganisms, and Louis Pasteur reported the development of vaccines for the prevention of fowl cholera and anthrax. Three years later, Koch set down his famous postulates which continue to serve as a guide in establishing the etiology of infectious diseases. Lord Joseph Lister, professor of surgery at the University of Glasgow and later at the University of Edinburgh, had already successfully applied the principles of Pasteur's germ theory of disease to surgery, and Koch had shown that various infectious diseases could be produced by injecting "putrid" fluids into animals. Establishment of Pasteur's theory and the perfection of bacteriological techniques by Koch—the use of agar in culture media and staining techniques and oil immersion for the microscopic study of bacteria—opened the way for the rapid advances that were made early in the twentieth century.[9,10] By the 1920s, bacteriology had become an important scientific discipline, and in time the term *microbiology* was introduced to connote the science that concerns viruses, protozoa, and fungi, as well as bacteria.[11-13]

Because Alexander Fleming's early years in medicine coincided with this period of proliferating activity in the isolation and characterization of species of microorganisms and their relation to infectious diseases, it is not surprising that he chose the field as his life's work.

5

PLATE 1.1. Sir Alexander Fleming and Sir Almroth Wright. (Reproduced with permission from: A. Maurois, The Life of Sir Alexander Fleming *[New York: E. P. Dutton, 1959], opp. p. 224.)*

Born on a farm in Ayrshire in 1881, Fleming had followed one of his brothers into the study of medicine, entering St. Mary's Medical School in 1901. He accepted a post as medical bacteriologist at St. Mary's on the day he completed his courses and in 1906 joined the staff of the Inoculation Department under the direction of Sir Almroth Wright. He qualified in surgery in 1908 and later became a Fellow of the Royal College of Surgery, but he never practiced as a surgeon to any great extent.[14]

Fleming spent his entire professional career at St. Mary's. Upon the retirement of Sir Almroth in 1945, he was named principal (director) of what had become by then the Wright-Fleming Institute of Microbiology. He continued there until his retirement on January 15, 1955. Fleming died two months later on March 11 and was buried in the crypt of St. Paul's Cathedral in London, an honor conferred on only a few of the most illustrious Britons. He had been knighted in 1944 and received the Nobel Prize for Medicine the following year.[15,16]

PLATE 1.2.   Lady Sareen Fleming. (Reproduced with permission from: A. Maurois, The Life of Sir Alexander Fleming *[New York: E.P. Dutton, 1959], opp. p. 97.)*

Fleming inherited from Sir Almroth a keen interest in the destruction of bacteria by leucocytes (white blood cells, also known as phagocytes). During World War I, he spent considerable time investigating problems associated with septic wounds. He was particularly impressed with the antibacterial power of the leucocytes contained in the pus exuded from wounds and also was impressed by the fact that the chemical antiseptics in common use were far more destructive to leucocytes than to bacteria. From his observations, he became convinced that the ideal antiseptic or antimicrobial agent must be highly active against microorganisms but harmless to leucocytes.[17,18]

7

Fleming's work with antiseptics and leucocytes continued after World War I, and in 1922 he discovered a substance he regarded as approaching the ideal antiseptic—lysozyme.[2] Two years later, he modified a technique first described by Sir Almroth in 1923, adapted it to the study of antiseptics,[19,20] and clearly established that only lysozyme (which was never of any therapeutic value) approached his concept of the ideal antiseptic. Later he showed with this technique that penicillin met the same criteria.

In September 1928, Fleming was growing strains of staphylococci (organisms that cause boils as well as more serious infections) and noticed one day a contaminating colony growing on one of his plates. The contaminant was obviously a mold, and it seemed probable that a spore from the air circulating in the laboratory had lit upon the plate. Around the contaminating colony, the staphylococcus colonies appeared transparent, as if they were being lysed or dissolved. Fleming recognized that anything capable of dissolving staphylococci might be biologically important, so he isolated the contaminant for further study. Later he said, "I must have had an idea that this was of some importance, for I preserved the original culture plate."[21]

In retrospect, we know that the contaminant must have been producing penicillin which diffused out from the colony and acted on the surrounding staphylococci. Today, the action of penicillin usually is more easily detected by observing its ability to inhibit growth of susceptible microorganisms. Fleming, however, first detected the substance by observing its lytic action on organisms that had already grown enough to produce colonies.

For many years after his discovery, numerous photographs, many of them taken by Fleming, were stored in a cabinet in his laboratory. Among the photographs was the original agar plate itself, showing the lysis of staphylococci in the vicinity of the contaminating mold (originally designated *Penicillium rubrum*, later correctly identified as a strain of *Penicillium notatum*) (see plate 1.3).[22] That plate and a replica are now located in the British Library (London), Collection 56209, vol. CIV, Fleming 1928 (z.3.c.1.). The original shows only a thin, well-dried shell of what was the culture medium. The staphylococcal colonies are visible around the contaminant and appear flat and transparent. There is no indication of any inhibition of growth around the contaminant. Rather, all colonies on the plate (except that of the contaminant) appear to have been lysed, indicating that the penicillin had acted on mature organisms.[1,23,24]

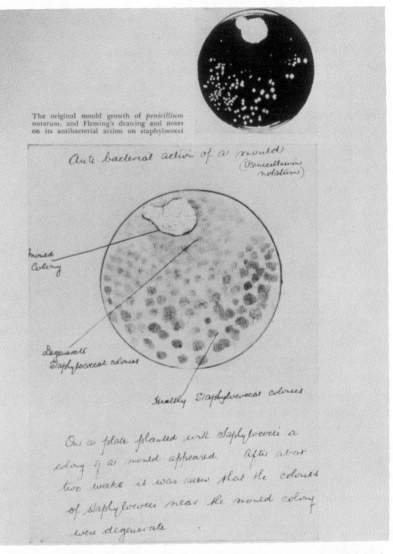

The original mould growth of *penicillium notatum*, and Fleming's drawing and notes on its antibacterial action on staphylococci

Anti bacterial action of a mould (*Penicillium notatum*)

mould Colony

Degenerate Staphylococcal colonies

Healthy Staphylococcal colonies

On a plate planted with Staphylococci a colony of a mould appeared. After about two weeks it was seen that the colonies of staphylococci near the mould colony were degenerate.

PLATE 1.3.   *The original mold growth of* Penicillium notatum *and Fleming's drawing and notes on its antibacterial action on staphylococci. (Reproduced with permission from: A. Maurois,* The Life of Sir Alexander Fleming *[New York: E. P. Dutton, 1959], opp. p. 192.)*

In contrast, the so-called replica shows around the contaminant a clear zone in which no growth occurred. This zone is surrounded by a second zone in which the staphylococcal colonies are smaller and fewer in number than in the area farther removed from the contaminant. Growth inhibition, but not lysis, is readily apparent.[23,24] Thus, the "replica" shows the growth-inhibitory action of penicillin, whereas the original plate shows penicillin's lytic action. (The difference between the "replica" and the original plate is unaccounted for.) Fleming apparently assumed, or knew, that the two effects were due to a single agent, for almost immediately he turned his attention to studies of its growth-inhibitory effects only.

Thus, Fleming's discovery of penicillin was not merely a fortuitous observation. It was a direct outgrowth of his discovery of lysozyme and his six years of study of its lytic action. If he had not already been interested in bacteriolysis, he might well in 1928 have discarded the contaminant that produced penicillin.

The Penicillin Record, 1928

Fleming's laboratory notebooks contain mostly records of patients seen and microorganisms isolated from their lesions, body fluids, or tissues. Few experiments are described in detail. Protocols are often presented without results, or results are cited without indicating the purpose of the experiment. Rarely is there a description of the methods used. Fleming worked on a small scale, did his own bench work, and was a meticulous and superb technician. Apparently he carried a mental record of what he had done and what should follow. Although incomplete, his records show his interest in the etiology of infection and in methods of destroying pathogenic, or disease-producing, microorganisms.

Fleming's notes do not describe the culture plate that first showed the action of penicillin. Although he may have made notes that have been lost, the detailed description of the plate in his first publication on penicillin (written eight to nine months after the discovery) seems to be the only available written record. On October 30, 1928, however, he did record in his notes an experiment that confirmed that the mold he had isolated as a contaminant produced a bacteriolytic substance active against staphylococci. Confirmation is as important as the initial observation.[25]

Fleming nevertheless knew the biological importance of the observation, for he showed the original plate to his associates at St. Mary's, preserved it, and subcultured the contaminant for future use. The original Fleming strain of *Penicillium notatum* was designated No. 4222 in the British National Collection of Type Cultures, located at the Lister Institute in London.[26,27]

Whether speaking or writing, Fleming never wasted words. His original manuscript on penicillin, for example, gives results and conclusions with only vague descriptions of the methods used to reach the conclusions. Ronald Hare, who knew Fleming personally for many years, recalls that "what mattered most to Fleming was not the recording of his experiments but their performance. . . . It was the actual performance of the experiment that he loved."[16]

The term *penicillin* first came into use on March 7, 1929.[23] All prior experiments mentioned in Fleming's notebooks were conducted with "mould juice," "mould filtrate," or "mould fluid." Initially the new term was applied to the penicillium culture fluids or "mould juices" with which Fleming worked. By common usage, however, *penicillin* came to refer to the active antibacterial substance(s) in the culture fluids. It is now used to refer specifically to certain antimicrobial substances derived as metabolic products of *Penicillium notatum* or *Penicillium chrysogenum*.

In the months immediately following his discovery, Fleming established that the antibacterial substance could be produced by certain strains of *Penicillium*, but not by all molds. He studied methods of producing it, established its presence in culture filtrates, determined its stability (lability) at various temperatures and its antimicrobial spectrum, and showed that it is nontoxic and nonirritating. He was most impressed with the fact that it did not interfere with leucocyte function.

On May 10, 1929, he submitted to the *British Journal of Experimental Pathology* his first report on penicillin and summarized his findings:

A certain type of penicillium produces in culture a powerful antibacterial substance. The antibacterial power of the culture reaches its maximum in about 7 days at 20° C and after 10 days diminishes until it has almost disappeared in 4 weeks.

The best medium found for the production of the antibacterial substance has been ordinary nutrient broth.

The active agent is readily filterable and the name "penicillin" has been given to the filtrates of broth cultures of the mould.

Penicillin loses most of its power after 10 to 14 days at room temperature but can be preserved longer by neutralization.

The active agent is not destroyed by boiling for a few minutes but in alkaline solution boiling for 1 hour markedly reduces the power. Autoclaving for 20 minutes

11

at 115° C practically destroys it. It is soluble in alcohol but insoluble in ether or chloroform. [In the salt form as in culture media, probably at pH 7–8, it was insoluble. Later, free penicillin was found to be highly soluble in ether and for a time ether was used in the solvent extraction of penicillin from fermentation liquors.]

The action is very marked on the pyogenic cocci and the diphtheria group of bacilli. Many bacteria are quite insensitive, e.g. the colityphoid group, the influenza-bacillus group, and the enterococcus.

Penicillin is non-toxic to animals in enormous doses and is non-irritant. It does not interfere with leucocytic function to a greater degree than does ordinary broth.

It is suggested that it may be an efficient antiseptic for application to, or injection into, areas infected with penicillin-sensitive microbes.

The use of penicillin on culture plates renders obvious many bacterial inhibitions which are not very evident in ordinary cultures.

Its value as an aid to the isolation of *B. influenzae* has been demonstrated.[1]

Thus, a remarkably extensive series of experiments was recorded and interpreted. Short of injecting penicillin into animals or conducting clinical trials in humans to determine its ability to prevent or cure infections due to susceptible microorganisms, he had done all that was necessary to establish its potential as a chemotherapeutic agent.[1,28] Although he failed to extract it from culture fluids, he showed that there were conditions under which penicillin would remain stable for considerable periods of time. It is interesting to note that his in vivo experiments were conducted with material probably containing not more than 1 to 2 units (0.6 to 1.2 mcg) of active penicillin per milliliter. An injection of 20 ml, described by him as a "huge" amount, probably contained no more than 10 to 20 mcg of active material.

Many have questioned why Fleming never tested the systemic chemotherapeutic activity of penicillin. It may be that he thought he had done so, for he believed strongly in the significance of the in vitro slide cell procedure he had developed. Moreover, his studies on chemical antiseptics and their use in the treatment of wounds (particularly during World War I) had conditioned him to think in terms only of topical therapy. More important, though, is the fact that in 1928–29, the worlds of basic research and applied research rarely conjoined. It was not considered important that basic research should lead to some practical application.

During the 1930s, Fleming received little credit for his discovery of penicillin. His associates in the Inoculation Department at St. Mary's continued to regard it as a substance of no real importance. Later they would credit the discovery to chance alone.

It is undeniable that the contaminating spore(s) lit upon the plate of staphylococci at just the time in their growth cycle when they were susceptible to the lytic action of the penicillin produced by the spores. As was shown in our laboratories in 1942 and has been shown repeatedly since, penicillin is active only when cell multiplication is taking place. Any decrease in the rate of growth of microorganisms results in a decrease in the rate at which penicillin acts. Lysis does not occur routinely; the conditions required for growth inhibition and for lysis are not identical.[29,30]

Nevertheless, Fleming was prepared for the fortuitous event. Because of his experience with lysozyme and the fact that lytic agents were much on his mind, he observed the contaminant, noted its lytic action, and preserved the plate. None of this was a matter of chance.

At the time of the discovery, Fleming was conducting a study of staphylococci and of mutants derived from them in preparation for a chapter he was writing for *A System of Bacteriology in Relation to Medicine*, to be published by the British Medical Research Council.[31] During the course of this study, he observed and isolated the contaminant and noted its lytic activity. Unfortunately, he failed to record the culture medium employed in his experiments, its pH, and other cultural conditions that might have contributed to the appearance of bacterial lysis. Moreover, although he preserved the contaminant, he failed to preserve the susceptible staphylococcal mutant.[32,33] In the light of recent studies, it seems possible that that strain of staphylococci may have been as important in its own way as the contaminant, for even now, after more than fifty years, there is inadequate understanding of the lytic phenomenon.[34–36]

Writing in 1979, Alexander Tomasz of the Rockefeller University commented:

The beta-lactams [which include all penicillins] undoubtedly represent the single most effective group of antimicrobial agents ever discovered, and the interaction of these antibiotics with live bacteria poses intriguing questions on many levels. What structural features of the beta-lactam molecule are responsible for the remarkably selective interference with murein metabolism: How do these drug molecules penetrate the outer layer(s) of the bacterial surface en route to their biochemical targets? What are the functions of the penicillin-binding proteins? Which are the penicillin-sensitive enzymes of murein metabolism and do penicillin molecules inhibit these? And, finally and most importantly, how does all this lead to the death and lysis of a bacterium? . . . Within a decade after the isolation of penicillin, . . . it

13

was noted that the mode of action of penicillin had several unique features not shared by most other antimicrobial agents: the selective toxicity of penicillins against bacteria in contrast to their virtual inertness against eukaryotic cells; penicillin-treated bacteria often underwent characteristic morphological changes; most bacteria rapidly lost their reproductive ability; and many species eventually lysed during penicillin treatment. Another peculiarity was an apparent concentration-dependent selectivity of action.... Most puzzling of all was the finding that the action of penicillin required that the bacteria be engaged in active growth (protein synthesis) at the time of addition of penicillin.... A clear distinction between the three types of antimicrobial effects of penicillins (inhibition of growth, killing of bacteria, lysis) was demonstrated [with some bacterial species].... The unique effectiveness of beta-lactams as antibacterial agents resides in the fact that the effects of these antibiotics are irreversible in most bacteria due to inactivation of the cells' reproductive capacity to the induction of physical destruction (lysis), or to both.... The complexity of the irreversible effects of penicillin may be better appreciated by recalling that the relationship between lysis and loss of viability is far from clear in any of the bacterial species studied. Although most bacteria killed by penicillin also undergo lysis, loss of viability invariably precedes lysis.[35]

Initially, Alexander Fleming observed, on his original plate, only lysis, the end stage of penicillin's action.

On balance, then, chance obviously contributed to the discovery of penicillin, but Fleming's experience, intuition, and keen powers of observation were all important to the discovery. As Ernst Chain commented in 1979 only a few months before his death on August 12:

Fleming made a very important and very original new biological contribution towards the discovery of the curative properties of penicillin. He was certainly very much favoured by good luck, but this is the case in most important discoveries.... There is no doubt that this discovery, which has changed the history of medicine, has justly earned him a position of immortality.[37]

2

Links—Conceptual and Personal—among Lysozyme, Other Mucinases, and Penicillin

Fleming studied lysozyme more intensively than he ever studied penicillin.[1-9] From the beginning, he recognized that its properties are similar to those of enzymes, and for this reason he named it lysozyme. He showed that it is a thermolabile substance, which is destroyed or in activated by heat and is present in tears, nasal secretions, leucocytes, and many tissues including skin; it is, however, absent from urine, cerebrospinal fluid, and sweat. He demonstrated its presence in some animals and plants and showed that it is highly active in egg white. He also isolated an organism that he designated *Micrococcus lysodeikticus* (i.e., capable of being dissolved) and showed that not only is it susceptible to lysozyme's action but it is particularly useful for assaying preparations of the enzyme. Finally, he observed that lysozyme is more active against nonpathogenic than pathogenic microorganisms. This observation led him to suggest that the greater an organism's sensitivity to lysozyme, the less its pathogenicity is likely to be. Most important of all, he postulated that lysozyme plays a role in the natural resistance of the body to infection.

Lysozyme (or muramidase) is a mucinase, an enzyme capable of hydrolyzing mucopolysaccharides (mucins) present in various secretions of the body. It achieves its bacteriolytic effect by hydrolyzing the 1,4-beta links between N-acetyl muramic acid and N-acetyl glucosamine, thereby destroying the cell walls of susceptible bacteria. Thus, Fleming's discovery of lysozyme, the outgrowth of his experimental investigations of antibacterial disinfecting agents, alerted him to the probable significance of bacteriolysis and paved the way for his later discovery of penicillin.

15

Fleming was not the only person who used lysozyme as a stepping stone to penicillin. Others who worked intensively with lysozyme before becoming interested in penicillin included Howard Florey at Cambridge University (later at Sheffield and Oxford universities) in England, Karl Meyer at the College of Physicians and Surgeons (Columbia University) in New York, and Ernst Chain and Edward Abraham at Oxford University.

Howard Florey was a pathologist with a keen interest in physiology. Lysozyme drew his attention in the late 1920s, while he was involved in studies on natural immunity. For some years, he had been interested in the action of mucin, mucous secretions, and the circulation of the blood and lymph. The presence of lysozyme in mucus and mucosal extracts of the stomach and intestines of several animal species made him certain of its biological significance.[10,11]

Florey was more interested in the possible etiological role of lysozyme in the development of duodenal ulcers than in its antibacterial properties. Ernst Chain, on the other hand, had worked with snake venom and had shown that the active principle in some venom is an enzyme, a complex organic substance from living cells that is capable of catalyzing certain chemical changes in organic substances within the cells. He suspected that lysozyme might also be an enzyme. Because he was interested in its antibacterial and bacteriolytic properties, he soon turned his attention to it.[12,13]

Meanwhile, Karl Meyer in the United States also initiated studies of lysozyme. Born in Germany as was Chain, Meyer acquired his M.D. and Ph.D. degrees in Cologne and Berlin before coming to the United States in 1930. He joined the Department of Biochemistry at Columbia's College of Physicians and Surgeons in 1933 and also became head of the biochemistry department of the Eye Institute of the Columbia University Presbyterian Hospital Medical Center. There he affiliated with Dr. Richard Thompson, a microbiologist, who suggested that they initiate studies on lysozyme. Fleming had shown that lysozyme is present in high concentration in tears, a readily available body fluid. Meyer, a chemist assigned to studies of diseases of the eye, had been at a loss for readily obtainable human tissue or fluids for analysis. With Richard Thompson and Drs. J. W. Palmer and Deborah Khorazzo as well, he published three reports in 1936 that established that lysozyme is an enzyme that initiates lysis by splitting certain polysaccharide components of the cell.[14–16] From vitreous humor he isolated a substance he named hyaluronic acid, a high-molecular-weight substance containing equimolar parts of glucosamine and glucuronic acid

PLATE 2.1.   Lord Howard Florey, M.A., M.D., Ph.D., F.R.S., Baron of
Adelaide and Marston (1898–1968). (Photograph by William H. Feldman, D.V.M.
Courtesy of the Mayo Foundation and Clinic, Rochester, Minn., and Lady Margaret
Florey.)

PLATE 2.2.   *Prof. Karl Meyer, M.D., Ph.D., in his laboratory at the College of Physicians and Surgeons, New York. (Photograph by Werner Wolff, from Black Star.)*

and an important component (as he showed later) of the ground substance of connective tissue.

Still another important advance was accomplished by Dr. A. E. H. Roberts, a biochemist who had moved in 1935 from the Dyson Perrins Laboratory at Oxford to the Sir William Dunn School of Pathology with which Chain was affiliated. He succeeded in purifying lysozyme in 1937.[17] Later, in 1939–40, Edward Abraham,[18] and L. A. Epstein and Chain[19] confirmed its carbohydrase nature and showed that lysozyme's substrate (a nitrogenous polysaccharide) exists in the cell walls of bacteria as well as in some tissues and body fluids. When acted on by lysozyme, the substrate breaks down into N-acetyl-glucosamine and a second substance that later was shown to be acetyl-muramic acid.

These were the first of many investigations on the structure of bacterial cell walls.[20,21] They were also the first in a long-continuing study by Meyer and others on the chemical nature of the ground substance of connective tissue and on the role of mucopolysaccharides in rheumatoid arthritis.[22-24]

Florey and Chain had achieved the goal of their studies on lysozyme. They had confirmed Meyer's observations on the mechanism of the action of lysozyme and, moreover, had contributed important information on the chemical nature of the substance. With the enzyme purified and the substrate identified, the mechanism by which it exerts its lytic effects seemed clear, and it was time for them to direct their attention elsewhere. Because they were both interested in naturally occurring antibacterial substances and in the phenomenon of microbial antagonism, they decided to undertake a systematic investigation of antibacterial substances produced by microorganisms. In an initial survey of the scientific literature on the subject, Chain came upon Fleming's original description of penicillin. Its lytic action aroused his and Florey's interest. If they were to study it, however, they needed financial support, so they submitted a funding application to the Natural Sciences Division of the Rockefeller Foundation, along with a review of their past scientific contributions and a list of proposed biochemical problems for future investigation.[25-27]

They suggested two studies: "(1) a chemical study of the phenomenon of bacterial antagonism with special consideration of bacteriolytic enzymes; and (2) a study of mucinases: their properties, mode of action, the chemical nature of their substrates, and their physiological or pathological significance."[25,27] In the section on bacterial antagonisms, penicillin was mentioned only once, and then indirectly, although they indicated that the substances produced by *Penicillium notatum*, *Pseudomonas pyocyaneus*, and *Bacillus subtilis* seemed most promising. They specifically mentioned gramicidin, an antibacterial substance shown by Dr. René Dubos at the Rockefeller Institute for Medical Research (now Rockefeller University) to be produced by cultures of a bacillus he had isolated from soil (*Bacillus brevis*).[28,29]

Florey and Chain commented often in later years that their interest had lain in conducting a purely academic study of microbial antagonisms, and said that it had never crossed their minds that one or more of these antibacterial substances might have chemotherapeutic potential.[13,21,25-27] Yet Chain had included in their funding application a proposal that they study systematically the chemical fundamentals of the phenomenon of microbial antagonism "with the aim of obtaining in purified state and suitable for intravenous injection bacteriolytic and bactericidal substances against various kinds of pathogenic microorganisms." He may well have

19

PLATE 2.3.   Prof. Sir Ernst Boris Chain, M.A., Ph.D., F.R.S. (1906–1979).
(Courtesy of Lady Ann Chain.)

had in mind animal studies of the type conducted earlier by Florey on the action of mucin, mucous secretions, and the circulation of the blood and lymph. He pointed out that a beginning had already been made at Oxford in purifying the bactericidal substances produced by *Penicillium notatum* and *B. pyocyaneus*.[25,27]

The study of mucinases proposed in Florey and Chain's application to the Rockefeller Foundation was rarely mentioned in later years. Professor F. Duran-Reynals, first at the Rockefeller Institute for Medical Research and later at the Yale University School of Medicine, published several reports between 1928 and 1933 that caught the attention of both Chain and Karl Meyer. Duran-Reynals described a "spreading factor" present in lysates and filtrates of cultures of invasive hemolytic streptococci, which increases the permeability of rabbits' skin to suspensions of India ink, some toxins, and bacteria.[30–32] In 1940, Epstein and Chain[19] observed that this substance is present also in testicular extracts. They showed that it hydrolyzes the mucopolysaccharide, hyaluronic acid, found by Karl Meyer and J. W. Palmer in 1936 in vitreous humour and the umbilical cord,[33] and by Forrest Kendall, Michael Heidelberger, and Martin Henry Dawson in 1937 in the capsular substance of Group A hemolytic streptococci.[34] Subsequently, in 1940, it became clear that the enzyme (hyaluronidase) that hydrolyses hyaluronic acid can be isolated from a number of bacteria—including some pneumococci, hemolytic streptococci, and anaerobic organisms—and from splenic tissue. It also became clear that hyaluronidase was probably very similar to Duran-Reynals's "spreading factor."[35–38] Not until some years later[39,40] was evidence presented suggesting that the increased permeability caused by hyaluronidase is not due merely to liquefaction or hydrolysis of the "ground substance" of connective tissue, but rather to the removal of the enzyme's substrate (i.e., to the removal of hyaluronic acid) from the ground substance.

A single report on hyaluronidase published by Ernst Chain and E. S. Duthie in 1940 is the only evidence of their continued interest in mucolytic enzymes.[41] Probably Florey, Chain, and their associates all assumed at the time that most, if not all, antibacterial substances were enzymes like lysozyme. They may also have assumed at the beginning that their proposed studies on mucinases and on antibacterial substances were closely linked. Whatever their assumptions then, however, the studies they initiated on penicillin must have quickly precluded their devoting time to other pursuits.

Florey and Chain's studies on the antibacterial agents produced by strains of *Pseudomonas pyocyaneus* and *Penicillium notatum* began almost a year prior to Great Britain's entry into World War II. The rise of Hitler during the 1930s and his aggressive policies, the failure of the Western countries' attempts at appeasement, Germany's takeover of Czechoslovakia in March 1939, and its ultimate invasion of Poland on September 1 of that year led England and France to declare war against Germany on September 3, 1939.[42] Ironically, it was on this same day—September 3—that the Third International Congress of Microbiology opened in New York City. Virtually all the world's leading microbiologists, including Alexander Fleming, were in attendance. But with their attention turned now to more pressing matters, the British contingent hastily arranged transportation home. Florey and Chain later insisted that the war in no way provided the initial impetus for their study of microbial antagonisms.[26,27,43] Nevertheless, it was on September 6 that Florey wrote to Sir Edward Mellanby of the British Medical Research Council (London):

Filtrates of certain strains of penicillium contain a bactericidal substance, called penicillin by its discoverer Fleming, which is especially effective against staphylococci, and acts also on pneumococci and streptococci. There exists no really effective substance acting against staphylococci, *in vivo*, and the properties of penicillin which are similar to those of lysozyme hold out promise of its finding a practical application in the treatment of staphylococcal infections.[44]

What is more, Florey and Chain's application to the Rockefeller Foundation was dated November 20, 1939.[27]

3

The Environment Surrounding
Early Efforts to Develop Penicillin,
1929–1939

Wright's Influence on Alexander Fleming

Alexander Fleming, as mentioned in Chapter 1, was strongly influenced by Sir Almroth Wright, perhaps more so than were others at St. Mary's Hospital. Sir Almroth was a strong exponent of vaccine therapy and sought to imbue the entire staff at St. Mary's with the same belief.

Before coming to St. Mary's in 1902, Sir Almroth had begun preventive inoculations against typhoid fever, using heat-killed cultures as vaccines. Subsequently, at St. Mary's, he studied other possibilities as the science of immunology began to evolve. Metchnikoff at the Pasteur Institute in Paris was developing his theories of phagocytosis, and Koch and others in Germany were preaching that the cause of immunity is the bactericidal power of the blood. Working with an assistant, Dr. Stewart R. Douglas, a fellow worker from 1902 until 1920 who succeeded him as director of the Biological Department at the National Institute for Medical Research in Hampstead (England), Sir Almroth soon showed that phagocytosis directly depends on the presence of some substance(s) in serum, and then demonstrated that the organisms had to be "prepared" by this substance before the phagocytes could engulf them. Sir Almroth called this component of serum an opsonin, from the Greek word *opsono* ("I prepare food for").

The importance of opsonins in immunity was established by Sir Almroth in a series of experiments showing that in persons infected with staphylococci or tubercle bacilli, the phagocytic power of their sera against these microorganisms was low, but could be specifically increased by active immunization. A technique was developed that was thought to measure a person's degree of immunity ("opsonic index" or "opsonic power"). For Sir Almroth this tecnique facilitated the development of specific homologous vaccines.[1-4]

Not all Sir Almroth Wright's beliefs proved correct as knowledge of immunology increased. Nonetheless, he stimulated several men, including Fleming, to work with him. He also had friends who ensured that funds for his department's research were available. Expenses at first were covered by contributions from private patients and friends who gave generously to the hospital. But in 1908, when the Inoculation Department of St. Mary's Hospital and Medical School was officially established, a committee, first chaired by Lord A. J. Balfour, was appointed to administer it. One of the committee's prime responsibilities was to forge a strong financial base for the department. To do this it was arranged that Parke Davis & Company, Ltd., would buy from the department and distribute to the public the vaccines prepared in the Inoculation Department Laboratories. The income from this arrangement was sufficient apparently for many years.[3,5]

Sir Almroth believed so strongly in vaccine therapy that some have said he was intolerant of anyone interested in other forms of therapy. Yet Leonard Colebrook wrote in 1954:

Just before the war [World War I] Paul Ehrlich (himself a great friend of Wright's) had startled the world by his brilliant discovery of "Salvarsan" and "Neosalvarsan" for syphilis. Wright had been intensely interested in this great new departure in medicine. He was indeed among the first to apply it in this country [England].

Soon after the war, Möllgaard and his colleagues in Denmark announced a similar success in tuberculosis—using a gold compound which they called "Sanocrysin." Although their triumph was shortlived, it stimulated Wright's mind to some fertile thinking about the principles which must underlie any successful chemotherapy.... His central theme was that in chemo- or pharmacotherapy it was essential to think, not only of the agent employed, but also of its transport, in effective amount, into the various infected parts of the body....

Fleming's classical experiments with *penicillium notatum*, which were destined to bear such a wonderful harvest in the hands of Florey and his team, interested Wright in so far as they were carried out in his Institute, but he did not foresee, any more than the rest of us, where they would lead. Nor was he greatly impressed by the discovery of prontosil and the sulphonamides.[3]

Whatever Sir Almroth's beliefs concerning the value of chemotherapy, two major contributions came from his department. Colebrook, a protégé and close friend who was also strongly influenced by Frederick Griffith, was a prime force in establishing the chemotherapeutic efficacy of the sulfonamides; and Alexander Fleming, then assistant director of the Inoculation Department, discovered penicillin.

Microbial Transformations in Pneumococci

Some of the most brilliant students of infectious diseases worked at St. Mary's Hospital during the 1920s and 1930s—Sir Almroth, Colebrook, E. W. Todd, Ronald Hare, and Fleming, among others. Although none of these men, other than Fleming, could ever claim a discovery of the magnitude of penicillin or lysozyme, each made contributions that earned him an important place in medical history. Besides the impact on their work of the teachings of Sir Almroth, all were influenced by Dr. Frederick Griffith, a bacteriologist and epidemiologist who served nearby as medical officer in the pathology laboratory of the British Ministry of Health. Frederick Griffith, through studies on microbial transformations in pneumococci, laid the groundwork for studies on immunological specificity and for the later discovery by others that deoxyribonucleic acid (DNA) is the substance that transmits hereditary characteristics—a discovery far more important even than the discovery of penicillin.[6-10]

In 1918, K. Baerthlein, in Germany, had reported the results of a detailed study of fourteen common bacterial species in which colonial types were correlated with pigment production, morphology of cells, fermentative power, virulence, and serological properties.[11] Following this, J. A. Arkwright (1921) observed a remarkable regularity in the appearance of two colonial types found among members of the colon-typhoid-dysentery group of microorganisms.[12] One of these was round, regular, and opague, with a glistening surface; the other was flat, irregular, and translucent, and had a granular surface. He designated the two as S type (smooth) and R type (rough), respectively. Arkwright also observed that the S type was readily converted into the R type on aging and reported that this transformation appeared to him to be irreversible.

In 1923, Frederick Griffith similarly observed two variant forms of pneumococci (*Diplococcus pneumoniae*, now designated *Streptococcus pneumoniae*), which he termed S and R.[6] Later (1928), he reported that

25

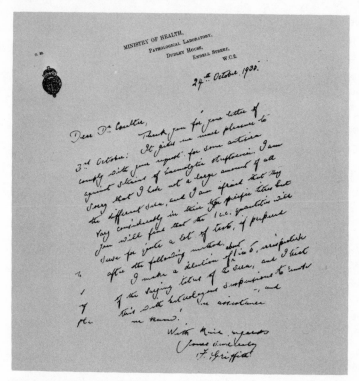

PLATE 3.1.   *Excerpt of letter from Dr. Frederick Griffith to Dr. Calvin Coulter, Columbia University, concerning the serological typing of hemolytic streptococci, October 24, 1935. (Courtesy of Dr. Florence M. Stone, State University of New York at Brooklyn.)*

when mice were injected with large numbers of living R type I or type II pneumococci, along with large numbers of heat-killed type III S pneumococci, it was possible to recover from the heart's blood type specific type III S organisms. He further showed that R forms of pneumococci, derived from any specific S type, could be transformed into S organisms of other types by subcutaneous injection, in mice, of small amounts of living R forms together with the killed vaccines of heterologous S cultures.[7]

Griffith's observation that one specific type of pneumococcus could be changed into another, through the intermediate R form, was confirmed by Dr. Martin Henry Dawson at the Rockefeller Institute for Medical Research in 1930, and in 1931, Dawson reported that similar results could be obtained in vitro by growing a small inoculum of R forms in media containing killed vaccines prepared from heterologous S cultures.[13,14]

PLATE 3.2.   Martin Henry Dawson, M.D. (Courtesy of Shirley Dawson Kirkland.)

Dawson's studies on pneumococcal transformations stemmed directly from those of Griffith and led to a continuing series of investigations conducted within Dr. Oswald Avery's department at the Rockefeller Institute over a period of many years, which clearly demonstrated the chemical nature of the transforming agent. Griffith had opened the door in 1928 and had pointed the way to what probably is the most important discovery in biology and medicine to date—namely, that deoxyribonucleic acid (DNA) is the component of bacterial cell genes responsible for microbial transformations.[15]

Griffith was not one to pursue only a single line of thought. He was a bacteriologist and an epidemiologist, widely interested in all infectious diseases and in their etiological agents. In 1934, he introduced a practical method for typing beta-hemolytic streptococci and thereby opened the way for epidemiological studies of a sort not possible previously.[16] In 1928, Rebecca Lancefield at the Rockefeller Institute for Medical Research had reported the classification of streptococci into a number of serological groups and presented evidence that the majority of strains isolated from lesions in man share a common polysaccharide. These she classified as Group A.[17-23] Subsequently, Griffith presented evidence that more than thirty serological types (the number is now more than fifty) exist among Group A strains.[24] (It is now recognized that two separate antigens are concerned in the type-specificity of the Group A strains—M antigen and T antigen, proteins soluble and insoluble, respectively, in alcohol.) In many instances, identification of the serological type to which strains of beta-hemolytic streptococci belong has allowed one to trace infections to their source(s).

Both Leonard Colebrook and Edgar Todd at St. Mary's Hospital Medical School were strongly influenced by these observations, as the nature of the chemotherapeutic studies they conducted during the 1930s makes clear.

Establishing the Basis of Chemotherapy

By 1930, the etiological agents of most infectious diseases that were then prevalent had been identified and characterized. Paul Ehrlich, a physician and director of the Institute for Experimental Therapy and of the Georg Speyer-Haus for Chemotherapeutic Research in Frankfurt, had for some time recognized the potential of chemotherapy for the treatment of bacterial infections. After he had made his beliefs known, he had been generously honored by development of these facilities for his research. Ehrlich's first experiments on infection had been performed in 1891 when, after observing that methylene blue would stain malaria parasites effectively, he and Dr. P. Guttman at the Moabit Municipal Hospital in Berlin tried the dye in the treatment of two malaria patients. The results were successful but not impressive enough to imply that methylene blue should be substituted for quinine which had long been in use.

As Ehrlich and Guttman reported:

This effect of the stain on the plasmodia in the first place, and then the knowledge that methylene blue, when infused into the blood of warm- and cold-blooded animals, stains certain inclusions of the red blood corpuscles—particularly, and with great uniformity, the nuclei of nucleated red blood corpuscles—gave us the idea of testing the therapeutic effects on malaria of methylene blue, the harmlessness and, on the other hand, the beneficial action of which we knew from many of our own observations. Our expectations have been completely realized. We can prove that methylene blue has a pronounced action on malaria. Under the administration of methylene blue, the attacks of fever cease in the course of the first few days and the plasmodia disappear from the blood after a week at the latest.

This action of methylene blue is a most striking one, when one considers that modern synthetic chemistry has been striving in vain to produce a remedy with a curative action on malaria.[25]

Ehrlich began to study chemotherapy intensively during 1902. By then, methods were available for production of many infectious diseases in animals. The in vitro bactericidal action of phenolic compounds and dyes was known, and the concept of bacterial immunity and the effectiveness of serum therapy had both been established. With Dr. K. Shiga, Ehrlich discovered that trypan red was curative and prophylactic in mice infected with *Trypanosoma equinum*,[26] thereby providing the first evidence that an experimentally produced disease could be cured by administering a synthetic organic substance of known chemical composition. It led him to an extensive study of other dyes and to the important observation that parasites can develop drug-resistance in vitro. For the first time, strains of trypanosomes were developed that were resistant to some compounds within a class of chemicals, yet were susceptible to other trypanocidal substances.

During this period, Ehrlich also studied the action of atoxyl against trypanosomes in vitro, found it ineffective, and therefore did not test it in vivo.[26] Soon after, however, it was shown by Thomas and Breinl at the

Liverpool Institute for Tropical Diseases[27] that the substance would cure trypanosomiasis in mice, and Robert Koch demonstrated its curative action on the human form of sleeping sickness. Ehrlich then reevaluated the chemical structure of atoxyl. After demonstrating that it was actually the sodium salt of para-aminophenyl-arsenic acid, he proceeded to study various homologues prepared by substitution in the amino group.[26]

In 1912, Ehrlich and Bertheim reported the synthesis of the hydrochloride of dihydroxy-diamino-arsenobenzene (serial No. 606 in their records), later known as salvarsan.[28,29] Salvarsan proved effective in curing relapsing fever in mice and in man, and Ehrlich and Hata showed its efficacy also in curing syphilis and trypanosomiasis.[30] Neosalvarsan, a more soluble derivative of salvarsan developed by Ehrlich in the same year, is a condensation product of salvarsan with sodium formaldehyde sulfoxylate and is a highly effective antisyphilitic agent. The discovery of salvarsan and neosalvarsan marked a turning point in medical history. They were the most important achievements in chemotherapy prior to the discovery of the sulfonamides.

Paul Ehrlich in 1908 shared with Élie Metchnikoff the Nobel Prize in physiology and medicine for studies in immunology. His lecture at the award ceremony, combined with one he presented five years later at the 17th International Congress of Medicine, summarized the results of more than twenty years of his studies on immunology and chemotherapy. By 1913, he had provided a "definite and sure foundation for the scientific principles and methods of chemotherapy," a base for all future work in the field. Working primarily with trypanosomes, he had established the specificity of effective chemotherapeutic agents, the significance of the development of resistance to them, and the importance of chemical structure in relation to activity. He had demonstrated, moreover, that drugs must be distributed in vivo to the actual sites of infecting agents and that appropriate dosages and combinations of chemicals must be used if they are to be chemotherapeutically effective. For these achievements, Ehrlich claimed no special credit, stating rather that "from the very beginning, chemotherapy has been in existence, since almost all the medicaments which we employ are chemicals." He shared the credit with all those who had simultaneously contributed to the fight against infectious diseases.[31,32] Paul Ehrlich nevertheless had established the basic principles of chemotherapy. He died in 1915 at the age of sixty-five. More than fifteen years passed before the next significant event occurred.

The Introduction of the Sulfonamides

In 1932, Gerhard Domagk observed the effectiveness of the hydrochloride of 4'-sulfonamido-2,4-diaminoazobenzene in the treatment of streptococcal infections in mice.[33] This was yet another important milestone in medical history.

The discovery of the chemotherapeutic efficacy of this compound, soon known as prontosil, aroused great interest. Never before had it been possible to cure a bacterial infection by administering a chemical agent. The magnitude of the discovery may be gauged by the fact that only ten years later the United States alone produced more than 10 million pounds of sulfonamide drugs in a single year (1943)—an amount estimated to be sufficient to treat more than 100 million patients.[34]

The history of the discovery of prontosil starts with the first synthesis of para-aminobenzenesulfonamide by P. Gelmo in 1908.[35] In the following year, Hörlein, Dressel, and Kothe of the I. G. Farbenindustrie in Wüppertal-Elberfeld, Germany, had prepared some dyes with sulfonamide and substituted sulfonamide groups. Their aim was to develop dyes for textile purposes.[36] It was five years later that Eisenberg[37] first noted in vitro the bactericidal activity of chrysoidine (2:4-diaminoazobenzene) and still later (1919) that Michael Heidelberger and W. A. Jacobs in the United States developed a number of azo dyes based on hydrocuprein and hydrocupreidine which also had strong bactericidal activity in vitro.[38] The preparation of the basic form of prontosil was postulated by the I. G. Farbenindustrie scientists in the English patent no. 149,428 in 1920 but was not fully described in the English literature until 1935.[39,40] Between the time of its postulation and its actual synthesis, interest in chemotherapeutic agents developed slowly.

Gerhard Domagk became director of research and experimental pathology and bacteriology at the I.G. Farbenindustrie in 1927. At the time, scientists were well aware of Ehrlich's discovery of the antisyphilitic activity of salvarsan [30] and of the observations of W. Schulemann and his associates[41] and of H. Mauss and F. Mietsch,[42] also at the I. G. Farbenindustrie laboratories, on the antimalarial activity of plasmoquine and atabrin. Dr. J. Morgenroth and R. Levy, moreover, had demonstrated not long before the curative effects of ethylhydrocuprein hydrochloride against experimental pneumococcal infections in mice.[43] In Germany, at least, chemotherapy was gaining credibility.

It was decided, therefore, at the I. G. Farbenindustrie to examine a large number of chemicals for their possible chemotherapeutic effects, following the procedures used so effectively by Paul Ehrlich. They would test all compounds in vivo, not merely in the test tube, since methods were not then available for the in vitro cultivation of some pathogenic microorganisms or for the development of many infectious diseases in experimental animals.[39,44] To test every compound in vivo, it often would be necessary to resort to trials in human beings.

During the course of this large-scale evaluation program, Domagk in 1932 became involved in studies on sulfonamide-containing azo dyes. He observed that the synthetic red dye, 4'-sulfonamido-2,4-diamino-azobenzene (later called prontosil), when administered orally to mice infected with lethal doses of beta-hemolytic streptococci, prevented the development of an otherwise fatal infection.[33,39,44] He also noted that the dye modified the course of staphylococcal infections in rabbits, but had no effect on pneumococcal or certain other infections experimentally produced in mice. The compound showed little or no activity in vitro. I. G. Farbenindustrie's decision to test all compounds in vivo, without regard for their demonstrable in vitro effects, had paid off as it had with Erhlich. The sulfonamides were the first class of drugs accepted as clinically useful in the treatment of bacterial infections.

Domagk did not publish his experimental studies on prontosil until 1935—three years after the first observations had been made.[33] What had transpired between 1932 and 1935 was the subject of much speculation for a time. Why did Domagk wait three years to announce his remarkable findings? Never before had there been any reason to believe that bacterial infections could be prevented or cured by the administration of chemicals. The information was too important to be withheld so long. As late as 1964, Colebrook, in a biographical memoir on Gerhard Domagk, suggested that it might have been difficult for him to confirm his initial observation, for by then it was known that all strains of a given microbial species are not equally susceptible to the action of antimicrobial drugs.[44] Surely this was not the case, however, for the observation was confirmed with ease by myself and many others promptly after Domagk's disclosure in 1935. One may speculate that Domagk was concerned about the fact that prontosil was effective only in vivo, not in vitro.

The events of the years between Domagk's first observations on prontosil and his report to the scientific community were recorded in detail by Dr. Perrin Long and Dr. Eleanor Bliss of the Johns Hopkins Medical

School in 1939.[39] Long had been present at the International Congress of Microbiology in London in the summer of 1936, met those who had been investigating the action of prontosil in Germany, and—while memories were fresh—recorded a full account of that memorable period.

Clinical confirmation of Domagk's experimental observations was reported simultaneously with the first report on his experimental observations.[45–47] It should be noted that initial reports on the clinical effectiveness of prontosil had appeared previously. In fact, the first announcement of the action of prontosil came not from Domagk but from a Dr. Foerster who on May 17, 1933, presented at the monthly meeting of the Düsseldorf dermatological society a report on the successful treatment of staphylococcal sepsis in a child with periporitis.[48]

Almost immediately after Domagk's announcement of his observations, Dr. Constantin Levaditi and Dr. A. Vaisman reported confirmation of the observations in France,[49] and later in the same year, Dr. F. Nitti and Dr. D. Bovet[50] noted variable chemotherapeutic results with 4'sulfonamide-2,4 diaminoazobenzene. Still later in that year, Dr. and Mrs. J. Tréfouël, with Nitti and Bovet at the Pasteur Institute in Paris, postulated that the drug underwent modification in vivo, breaking down at the azo linkage to form para-aminobenzenesulfonamide, a substance shown by Prof. E. Fourneau also at the Pasteur Institute to be as effective as prontosil itself in the treatment of animals experimentally infected with hemolytic streptococci.[39,51] In June 1936, Leonard Colebrook and Maeve Kenny in England published an extensive report confirming the German experimental and clinical observations. They showed that prontosil increased the bacteriostatic power of the blood of patients, and provided evidence conservatively suggesting that it might be of value in the treatment of human puerperal sepsis:

There have been 3 deaths in the 38 cases treated by prontosil—that is 8 per cent. In considering the significance of this figure it has to be borne in mind that the first 10 cases were chosen because they were severe or moderately severe cases—the mildest being deliberately excluded; and, further, that the death rate fluctuates to some extent according to whether few or many late cases—particularly late peritonitis cases—are admitted to the hospital. The following figures give an idea of the usual mortality-rate for all cases admitted to the isolation block and found to be infected by haemolytic streptococci.

In 38 cases admitted immediately prior to the use of prontosil...there were 10 deaths—e.g., 26.3 per cent. In the 38 cases immediately preceding these there were 9 deaths—i.e., 23.7 per cent. During the four years 1931–1934 the rate varied between 18 and 28.8 per cent (average 22 per cent).

While, therefore, there would appear to have been a very considerable reduction of the death rate among the prontosil-treated cases it would be unwise to assume on the basis of so small a series that the reduction will be maintained. Nevertheless we are of the opinion that the very low death-rate, taken together with the spectacular remission of fever and symptoms observed in so many of the cases, does suggest that the drug has exerted a beneficial effect. It may be added that there is no reason to ascribe the clinical results to any other change in the local or general treatment of the cases.[52]

In 1936, Buttle, Gray, and Stephenson[53] also confirmed the observations made in France that showed that para-aminobenzenesulfonamide (sulfanilamide) had a curative effect on streptococcal infections in mice; they showed too that it was effective against meningococcal infections. Finally in 1937, Fuller (in England) performed the critical experiments and demonstrated that prontosil—as Tréfouël and his associates had postulated—actually does break down in vivo, with the formation of para-aminobenzenesulfonamide.[54]

The observations of Colebrook and Kenny reported in 1936[52] were confirmed in 1939 in a larger series of patients by Colebrook and Purdie.[55] All doubts concerning the merits of chemotherapy were laid to rest.

Sulfapyridine, synthesized by Dr. A. J. Ewins and his associates at May & Baker and initially designated M&B 693, was introduced in 1937,[56,57] and Dr. L. E. H. Whitby soon showed that this new chemical could exert a remarkable chemotherapeutic effect against acute pneumococcal infections in humans, particularly acute pneumococcal (lobar) pneumonia.[58] The therapeutic effects of sulfapyridine were so striking, in fact, that type-specific antisera, recently developed by the Lederle Laboratories for the treatment of acute pneumococcal infections, promptly became historical curiosity.

Prior to this time, few had thought constructively about the possible use of chemicals for treatment of bacterial infections. The early observations of Emmerich and Low with pyocyanase[59] and of Morgenroth and Levy with optochin[43] had stirred some interest but not enough; prejudice remained. Sulfapyridine, however, produced such startling therapeutic effects that it could not be ignored. Thus, it ran the gauntlet of opposition—well ahead of penicillin—and established that chemotherapy had more to offer than specific immunotherapy. It opened the door to new chemotherapeutic agents.

Sulfathiazole, also synthesized by Ewins and his team at May & Baker (in 1938), sulfadiazine, sulfamethazine, and sulfamerazine followed in quick succession.[56,57] Remarkable studies on the mode of action of the sulfonamides, on their absorption, excretion, and distribution in humans, were soon reported—studies that introduced the principles that have been followed in all subsequent investigations of antimicrobial drugs. These studies provided the basis for modern biochemical pharmacology, pharmacokinetics, and pharmacogenetics.[60–65] Other studies on the sulfonamides introduced the concept of metabolic antagonists[66–68,57] and led to the institution of regulations relating to drug safety. The groundwork for all that followed with the introduction of penicillin and the later series of antiinfectious drugs—the basic principles of chemotherapy—was established through research on the action of the sulfonamides.

The synthesis of hundreds—even thousands—of related compounds was unleashed by the studies on the sulfonamides. Probably no other class of compounds has yielded so many different and active agents as the sulfonamides. Even today, more than forty years later, the sulfonamides play an important role in the control of acute infectious diseases, particularly in developing countries where medically trained personnel are scarce and costs, ease of delivery, and safety of drug administration make the difference between success and failure in infection control.

Microbial Antagonisms and the Introduction of Antibiotics

While the sulfonamides were being developed and their efficacy demonstrated, thus establishing chemotherapy as a legitimate and effective means of treating infection, studies on microbial antagonisms in nature assumed increasing importance. The antagonistic relationships among microorganisms had attracted attention since the earliest days of bacteriology, and by 1939, the extensive studies of Maurice Welsch in Belgium, Selman A. Waksman in the United States, and others had made it clear that many substances are produced by microorganisms in nature that inhibit the growth of other organisms living in association.[69,70]

Of prime importance were the observations of René Dubos at the Rockefeller Institute for Medical Research. Dubos isolated from soil that had been enriched with various living bacteria a gram-negative, spore-bearing bacillus, *B. brevis*, which was capable of exerting a lytic effect against both staphylococci and pneumococci. From the culture medium, he isolated a substance he designated tyrothricin and which he and Dr. R. D.

35

Hotchkiss, also at the Rockefeller Institute, soon showed could be separated into two crystalline preparations, gramicidin and tyrocidin, both with antimicrobial properties.[71,72]

Gramicidin was ultimately introduced into clinical medicine—the first antibiotic (a substance produced by a microorganism and capable of inhibiting growth of certain other microorganisms) to be used in the treatment of infections in human beings.

Alexander Fleming: The Years from 1910 to 1930

Soon after the beginning of World War I, Sir Almroth Wright was asked to establish a laboratory and research center at Boulogne-sur-Mer in France. This was the first recognition that research could contribute to wartime medicine, a fact firmly established before the start of World War II. When Wright left for France, he took with him three of his staff, one of whom was Alexander Fleming. In Boulogne, Fleming's ingenuity and imagination contributed much to the development of necessary laboratory equipment and to the design of critical experiments. Fleming carefully studied infected wounds, noting both the importance of removing necrotic tissue as soon as possible and the fact that better results were often obtained if no antiseptics at all were used. During this period (1914–18), he studied intensively a large number of antiseptics and laid the groundwork for his future studies of these substances.[73–75]

Between 1929 and 1939, only two papers on penicillin, written by Fleming, appeared in the scientific literature. One, coauthored by I. H. Maclean, described the use of active broths in differential culture media.[76] The other concerned the action of penicillin and potassium tellurite in differential media.[77] In the latter report, Fleming listed five uses of penicillin, the first four of which related specifically to its use in culture media for microbiological purposes. The fifth was "as a dressing for septic wounds":

In penicillin we have a perfectly innocuous fluid which is capable of inhibiting the growth of the pyogenic cocci [i.e., of staphylococci] in dilutions up to 1 in 800. It has been used on a number of indolent septic wounds and has certainly appeared to be superior to dressings containing potent chemicals. It is unlikely that it acts by killing the bacteria directly. . . . The practical difficulty in the use of penicillin for dressings of septic wounds is the amount of trouble necessary for its preparation and the difficulty of maintaining its potency for more than a few weeks.[77]

Stuart Craddock, who worked in Fleming's laboratory at the time, published almost nothing on penicillin. He was a research scholar, only twenty-five years old and a trainee in the laboratory. Craddock's records indicate that they worked constantly with the mold and its antibacterial growth product(s).[78] His records are cryptic. One or more days' work often is described merely with such phrases as "extraction of the mould's inhibitor," "injection into rabbit of mould juice," "1st test if mould juice combines with staphylococci," "more trials at extraction of the active inhibitor," or "1st (probably) titration in serum." But three experiments

*PLATE 3.3.   S. R. Craddock, about 1929, St. Mary's Hospital, London.
(Reproduced with permission from: Ronald Hare,* The Birth of Penicillin *[London: George Allen & Unwin, 1970].)*

37

apparently seemed important enough to warrant their being recorded in more detail.[78]

On February 14, 1929, Craddock noted that his colleague at St. Mary's Hospital, Frederick Ridley, had extracted the active principle of "mould juice" with ether and had succeeded in concentrating the substance more than twentyfold. A week later on February 20, Ridley extracted the color from the mold juice by shaking with ether and salting out with potassium salts. He concluded that the yellow substance was not the active inhibitor—a fact confirmed somewhat later by Clutterbuck, Lovell, and Raistrick.[78,79]

In March 1929, Craddock described the injection of mold juice into a rabbit and, from this experiment, concluded that the "inhibitor does not remain free in serum for very many minutes." On April 2, he recorded that a rabbit was killed and its organs removed, placed in broth, and then inoculated with staphylococci. After incubation for twenty-four hours, the organs were placed in mold juice, again incubated, and then subcultured. From the experiment, Craddock concluded that mold juice could not penetrate the organ tissues, and he raised the question: "Does the juice require body temperature to act?" Whether it occurred to Craddock, or to Fleming, that the active principle in the mold juice could more easily penetrate these organs through the circulatory system if it were injected parenterally, the record does not indicate (see table 3.1).[78]

Many of the experiments described by Craddock in his laboratory notebooks are reported in Fleming's first paper on penicillin published in 1929.[80] Fleming's handwriting appears on one occasion at least in the middle of Craddock's records. The two undoubtedly worked very closely together, Fleming perhaps conveying verbally to Craddock what he thought should be done, Craddock maintaining the records of what actually was done. Certain of the experiments they performed suggest that the possibility of penicillin's having real chemotherapeutic activity crossed their minds often (see table 3.1). The available records indicate, however, that they may not have been too familiar with the techniques necessary for in vivo drug evaluation.

Fleming was the sole author of the first publication on penicillin; credit is given to Craddock and Ridley only through a brief acknowledgment: "In conclusion, my thanks are due to my colleagues, Mr. Ridley and Mr. Craddock, for their help in carrying out some of the experiments described in this paper, and to our mycologist, Mr. La Touche, for his suggestions as to the identity of the penicillium."[80]

TABLE 3.1   Penicillin Highlights: The Fleming Era

1928	
The Discovery	Fleming's Laboratory St. Mary's Hospital, Paddington

Alexander Fleming
sole discoverer
of penicillin

September	**October to December**
Alexander Fleming observes antistaphylococcal (lytic) activity of contaminating mold, at St. Mary's Hospital, London	Titles of experiments indicate Fleming's and Craddock's line of thought
October	"Extraction of mold's active inhibitor"
Lytic action of "mold juice" from growth of the contaminant confirmed by Fleming	"Test if mold juice combined with staphylococci"
Mold tentatively identified by Dr. C. J. La Touche at St. Mary's Hospital, as probable strain of *Penicillium rubrum*	"More trials at extraction of the active inhibitor"
	"Injection into rabbit and first (probably) titration in serum"
	"Effects of heat and acid on mold juice activity; effect of active inhibitor on time of bacterial killing, polymorphonuclear leucocyte function, etc."
	"Effect of mold extract on bacterial growth—filtered extract of inhibitor mold used"

Source: See note no. 23, chapter 1.

(*Continues*)

TABLE 3.1   Penicillin Highlights: The Fleming Era (*Continued*)

1929

The Notebooks of Stuart Craddock,
Research Scholar, in Fleming's Laboratory

January

First study of penicillin's effect on isolation of Pfeiffer's bacillus and nasal infections

February

Dr. Frederick Ridley, an ophthalmologist, joins Fleming and Craddock in study of active inhibitor in mold juice.

6th: Active inhibitor extracted into ether, distillation in vacuo at 37°C.

14th: Active substance extracted and concentrated twentyfold.

20th: Color extracted with ether, salted out with potassium salts. No inhibition by extract. "This would appear to show that the yellow substance is not the inhibitor."

March

12th: Active principle extracted with acetone. Inactive resinous deposit left behind.

March (*Continued*)

13th: "Active principle extracted more effectively with alcohol."

14th: "In spite of the Fleming mold having been planted 9 days, it still improves. This is unlike any other."

15th: "Experiments continued with Ridley."

29th: "The name penicillin has been given to the filtrates of broth cultures of the mold."

April

19th: "Five liter flasks" were planted.

May

10th: First report on penicillin submitted to the British *Journal of Experimental Pathology*

June
to
July

Experiments performed to prove if there is an enzyme that causes the mold juice to deteriorate.

Source: See note no. 78, chapter 3.

(*Continues*)

TABLE 3.1   Penicillin Highlights: The Fleming Era (*Continued*)

1930–1938

The Interim Years

1932: Clutterbuck, Lovell, and Raistrick

Confirm the production of active inhibitor by Fleming's mold

Confirm Ridley's observations on solubility of active inhibitor in ether

Isolate the pigment (chrysogenin) and

Confirm Ridley's observation that the pigment and the inhibitor are not the same substance

1930 to 1932: First patients, mostly "old sinuses," treated with mold juice by Fleming—with favorable results

1932: Fleming again calls attention to therapeutic possibilities of penicillin.

1932: Dr. C. J. Paine of Sheffield, former student of Fleming, successfully treats 3 of 4 patients with ophthalmia neonatorum, and 1 eye injury due to pneumococci.

1934: Dr. Lewis B. Holt, a chemist, joins the staff of the Inoculation Department at St. Mary's Hospital and, at the request of Alexander Fleming, attempts isolation of active growth inhibitor.

"Used amyl acetate (which does not mix with water) to extract the active inhibitor from acidified (pH 5–6) mold juice. Then back-extracted into a weak solution of sodium bicarbonate in water at pH 8.0."

1935: Dr. Roger Reid, microbiologist in the United States, unaware of Holt's observations, confirms Clutterbuck, Lovell, and Raistrick's observations.

PLATE 3.4. Upper left: photograph of a colony of penicillium and its inhibiting effect on the growth of various microorganisms. Below: Fleming printing photographs of bacterial culture plates in the darkroom of his laboratory. (Reproduced from A. Maurois, The Life of Sir Alexander Fleming [New York: E. P. Dutton, 1959].)

PLATE 3.5.   Frederick Ridley, about 1929, St. Mary's Hospital, London.
(Reproduced with permission from: Ronald Hare, The Birth of Penicillin *[London: George Allen & Unwin, 1970].)*

The record of Ridley's experiments on the extraction of penicillin into ether, later confirmed by Clutterbuck, Lovell, and Raistrick, appears only as a brief note in Stuart Craddock's notebooks. Frederick Ridley, unfortunately, never published his results. Thus, he lost the credit that it now seems he should have received. His failure to publish may have been due to the fact that the small amounts of penicillin produced precluded repetition of experiments. It was not customary in 1928–29 to publish the results of isolated experiments.

43

Fleming was more bacteriologist than clinician, despite his early training and place of work. He was a biologist, not a chemist; and a bench worker, not an administrator or team organizer. Most important, he was limited by the technology available to him. Sophisticated methods had not yet been developed for characterization and determination of organic compounds. Even Ernst Chain, skilled chemist that he was, later commented that without partition and paper chromatography, infrared and ultraviolet spectroscopy, nuclear magnetic resonance and electronic spin resonance—techniques not known in the 1920s and early 1930s—isolation of an unstable natural product such as penicillin would have been exceedingly difficult.[81–85] Admittedly, it is unlikely that Fleming could have utilized these very specialized techniques, even had they been available to him. But they undoubtedly would have been used by Prof. Harold Raistrick, a well-known biochemist at the London School of Hygiene and Tropical Medicine.

Penicillin in the 1930s

Harold Raistrick was the first person outside Fleming's laboratory to evince an interest in penicillin and was one of the very few who experimented with the substance prior to Florey and Chain's studies at Oxford in 1938. Raistrick was a natural products organic chemist, expert in the area of biosynthesis; he had been concerned for many years with mold products, particularly those from penicillia. Fleming's "substance" quite naturally sparked his interest.[79]

Raistrick obtained his culture for study directly from Fleming and the Lister Institute. Working with Reginald Lovell, a bacteriologist, and with P. W. Clutterbuck, a young biochemist, he cultivated the organism in a semisynthetic medium containing salts and glucose (the Czapek-Dox medium), which was later used by many others to produce penicillin. Raistrick and his associates succeeded in isolating the pigment (chrysogenin), which was produced with the antibacterial substance during growth of the mold and gave the mold filtrates their color. As Frederick Ridley had done earlier, they extracted the active substance with ether but were unable to recover it from the ether. Most of the activity was lost during evaporation of the solvent into which the penicillin had passed during their attempts at extraction.[79] Although not recorded by Craddock, the same may well have been true when Ridley tried to recover the penicillin from his high-potency ether extract.

PLATE 3.6.   Prof. Harold Raistrick at his desk, London School of Hygiene and Tropical Medicine. (Courtesy of Dr. Robert Coghill.)

It is said that Raistrick unsuccessfully attempted to interest some physicians in using the preparations of penicillin he and his associates made. It is said too that the major bacteriologist in England at that time, Prof. W. W. C. Topley, could see no future in penicillin and discouraged Raistrick's continuing his research on the substance. In any event, the study was soon discontinued.

As we have seen, Fleming knew from the beginning that his contaminating mold belonged to the genus *Penicillium*; it was for this reason he named the substance penicillin. He erred, however, in identifying the species. He first thought that it was a strain of *Penicillium rubrum*. His mycologist, C. J. La Touche, had suggested this possibility. The strain was so designated until Clutterbuck, working at the London School of Hygiene and Tropical Medicine with Harold Raistrick, began to wonder about the

45

color of the pigment. He asked Dr. Charles Thom, a well-known mycologist in the U.S. Department of Agriculture, to study the strain and fully identify it. Thom confirmed Clutterbuck's suspicion that it had been erroneously classified and reported it to be a strain of *Penicillium notatum* Westling in the *Penicillium chrysogenum* Thom series.[86–88]

Other than Raistrick and his associates in London, only Roger Reid at Pennsylvania State College (now Pennsylvania State University) in the United States showed interest in penicillin. Reid published in 1935 the results of an extensive study of a number of molds he had examined to determine which, if any, would produce Fleming's antibacterial substance.[89,90] Using Fleming's strain of *Penicillium notatum* as classified by Thom, he showed that the organism could be grown either in a veal infusion or in a synthetic medium. He also demonstrated its ability to produce an active antibacterial substance and confirmed Fleming's observations on the substance's spectrum of activity. In his experience, penicillin was bacteriostatic, not bactericidal. It inhibited growth of susceptible microorganisms, moreover, but did not lyse them. The substance was extremely unstable, as Fleming and his associates, as well as Clutterbuck, Lovell, and Raistrick, had shown earlier. It was also difficult to separate from the culture fluid. Using dialysis, absorption, and distillation at low temperatures, Reid was unable to separate out the penicillin.

Reid, who was a graduate student at the time, had no research funds and was responsible to a professor who, Reid himself related, saw no future in penicillin and thus considered it unsuitable for a doctoral dissertation.[91] Although Reid's studies on penicillin, therefore, were soon terminated, he had confirmed the observations of Fleming, Craddock and Ridley, and Clutterbuck, Lovell, and Raistrick. He had failed, however, to add new information concerning the isolation and purification of penicillin.

An attempt to do this was made by Lewis Holt, a chemist who joined the staff of the Inoculation Department at St. Mary's in 1934.[92,93] He was hired to assist in the preparation of antitoxins, but before long Fleming suggested to him that he might like to try purifying penicillin. Others in the department had tried and failed, but Holt, knowing Fleming's interest in the substance, was anxious to undertake isolation of the substance. It is said that he was given a copy of Clutterbuck, Lovell, and Raistrick's 1935 publication on penicillin but was not told of Ridley and Craddock's earlier attempts. This seems doubtful, but it probably would not have helped Holt to know of Ridley and Craddock's apparent failures.

Holt, however, did make one very important contribution. He learned from Clutterbuck, Lovell, and Raistrick's work that penicillin is soluble in organic solvents, and he used amyl acetate for extraction of the active material. Since amyl acetate does not mix with water, he added it to the mold juice at a pH of 5–6, shook the mixture well, and then separated the amyl acetate from the broth in a separatory funnel. Thereafter, he again thoroughly mixed it, this time with a weak solution of sodium bicarbonate in water at a pH of about 8. Some of the penicillin went into solution in the bicarbonate. The loss was great, for the bicarbonate was too alkaline. But Holt had shown that the active substance could be recovered from organic solvents by extraction back into an aqueous solution at an appropriate pH —a procedure suggested many years later by Dr. Norman Heatley and used with success at Oxford. Actually this procedure was not novel; it had been used for some time previously by those involved in the isolation and characterization of alkaloids from natural products.

Lewis Holt failed to recognize the importance of his contribution to the isolation and purification of penicillin. Considering his results unsuccessful, he lost interest, discontinued the study, and never published (see table 3.1).[92] From then until 1940, when Chain, Florey, and their associates at Oxford published their first report in the British scientific journal *Lancet* indicating their successful isolation of the substance and its obvious chemotherapeutic potential, only one other report appeared. It was written by Dr. S. Bornstein, an unknown investigator, who, four months before publication of Chain and Florey's report, published the results of experiments he conducted at the bacteriology laboratories of the Beth Israel Hospital in New York City.[94]

Bornstein had produced filtrates containing penicillin by growing Fleming's mold and had shown their usefulness in the classification of streptococci, particularly enterococci. But like others before him, he apparently was deterred from the use of penicillin by the instability of the substance. In any event, he presented no evidence that the possibility of using penicillin therapeutically had occurred to him. Four months later, the first publication by Florey, Chain, and their associates reached New York City and presumably came to Bornstein's attention.

Harold Raistrick later claimed that he had failed to isolate penicillin because he lacked "complete bacteriological cooperation." Fleming, moreover, claimed that he had lacked both bacteriological and clinical cooperation. The remarkable fact that 400- to 800-fold dilutions of Fleming's fermentation liquors—which we now know could have contained no more than 1 to 2 units of active penicillin per milliliter of liquor—effectively inhibited microbial growth was ignored. Fleming's discovery of penicillin did not arouse interest—not even enough to lead to its study by those closest to him at St. Mary's.

In retrospect, it is clear that these men lacked adequate space and personnel to grow the organisms in sufficient volume, that they lacked appropriate analytical techniques, and that they failed to realize the very minute quantities of the active material, on a weight basis, they were trying to isolate. Unquestionably too—as became apparent in later years—an interdisciplinary approach was needed for the development of penicillin. No single investigator, or small group of investigators, could have accomplished the task.

Early Attempts to Use Penicillin Clinically

This question has been asked repeatedly in the years since 1928: to what extent did Alexander Fleming utilize penicillin as a therapeutic agent between the time he first reported its existence and the time of Chain, Florey, and their associates' first description of its therapeutic efficacy? Judging from the written records, the answer is virtually not at all. Craddock's notebooks contain a page, undated but presumably written in March 1929, on which he comments: "Searching for a patient to treat?" The books contain no other references to penicillin's clinical use.[78] Nevertheless, despite the failure to isolate the substance, some attempts were made to evaluate its potential clinical role.

As mentioned previously, Fleming alluded in his first report on penicillin to its possible use on dressings for septic wounds and remarked that experiments to evaluate its potential in the treatment of pyogenic infections were in progress. In 1932, he further stated that it had been used on a number of indolent septic wounds and had appeared superior to dressings containing some other antiseptics. Fleming's original report is a classic: it describes all the important properties of penicillin and it indicates that Fleming knew that penicillin was more toxic to bacteria than to leucocytes and other animal cells. He knew from the beginning that penicillin was the ideal antiseptic he had sought for many years, but the "trouble" of making it seems to have dominated his thinking.[78]

About 1932, while Raistrick was attempting to isolate the active substance, Dr. C. G. Paine at the Royal Infirmary in Sheffield, who, like so many others, had been a pupil of Fleming's, decided to try growing

Fleming's strain of *Penicillium* in broth.[95] Apparently to his surprise, he produced varying quantities of the active substance. He applied a filtrate of the mold culture to three patients with staphylococcal infections of the skin, but without success. He then used it on four babies with eye infections (ophthalmia neonatorum). Two, whose infections were due to strains of *N. gonorrhoeae* (gonococci), were promptly cured. One of the other two babies, whose infections were due to staphylococci, responded also. Paine then used penicillin filtrate to treat a colliery manager. A small piece of stone had penetrated the man's right eye and lodged itself partly behind the pupil. The conjunctiva was inflamed, and the eyelids were badly swollen and distorted. Swabs taken from the conjunctival sac gave pure cultures of pneumococci, which Paine knew were susceptible to penicillin. He washed the man's eye with his filtrate at regular intervals over a period of two days; the organisms disappeared, and the piece of stone was then removed. The penicillin filtrate, crude as it must have been, had saved the man's sight.

Later Paine remarked, "The variability of the strain of *Penicillium*, and my transfer to a different line of work, led me to neglect further investigation of the possibilities of penicillin, an omission which, as you may well imagine, I have often regretted since."[95] Paine left the Royal Infirmary in Sheffield to work at the Isolation Hospital attached to the Jessop Hospital for Women. Although he saw many patients with puerperal sepsis, he made no attempt to use penicillin for their treatment.

The first patient treated with penicillin was Stuart Craddock who, from the beginning, had worked intensively with the substance in Fleming's laboratory. Craddock's laboratory notebook records that patient S.C. had an infection of the nose from which staphylococci, diphtheroids, Pfeiffer's bacilli (organisms that probably now would be designated "Hemophilus") and some other organisms were isolated. He "washed" the apparently infected nasal area with mold juice and recorded after three hours a decrease in the number of staphylococci but a beneficial effect on the growth of the Pfeiffer bacilli. No mention is made of the effect of the mold juice on the infection otherwise.[78] Ronald Hare, who was affiliated at St. Mary's Hospital at the time, has recorded that Craddock's antrum (presumably sinus) had become infected with a diphtheroid organism that was susceptible to penicillin. He irrigated the antrum twice daily with mold juice, but without effect. Shortly thereafter, the mold juice was used at St. Mary's for treatment of an infected amputation stump. Again the results were negative. Perhaps the first "cure" produced by penicillin was that of a man named Rogers, Craddock's assistant. Rogers developed an eye infection due to pneumococci, generally highly susceptible to penicillin. Treatment of the eye with mold juice is said to have led to the disappearance of the infection.[96]

Alexander Fleming's interest in penicillin never wavered. One wonders what the outcome would have been if he had been more aggressive, if he had believed more strongly in the observations he made in 1928 and 1929, if he had more strongly insisted that there must be conditions under which penicillin would be more stable—conditions under which the substance could be produced with greater ease and could be used more effectively over longer periods of time. One wonders too what would have happened if he had recognized the significance of Lewis Holt's observation on recovery of the active substance from organic solvents. However, Fleming, although said to be stubborn and determined at times, was not an aggressive person, and he was caught in a circular situation. Without enough penicillin of adequate purity to prove its chemotherapeutic efficacy, he was unable to arouse interest in studies on methods for its isolation and purification. Without the methodology, there was no possibility of obtaining penicillin suitable for clinical use. Despite all this, when a product pure enough for treatment of infection became available at Oxford more than ten years later, Fleming was quick to react. On the morning following his receipt of the August 24, 1940, issue of *Lancet*, describing the isolation of penicillin, he made his first and only visit to Florey's laboratories at Oxford to learn what had been done with his penicillin and to obtain a sample.[5,97,98] But not until 1942 did Fleming himself treat a patient who showed with absolute certainty the substance's therapeutic value.[99]

Infectious Disease in 1939

Fleming's interest in staphylococci was aroused at a time when these microorganisms were usually considered common and unexciting. Most of the staff at St. Mary's were not interested in them, feeling that their relatively low degree of pathogenicity made them unworthy of extensive research. Nevertheless, Fleming's studies on them led to his discovery of penicillin.

While he was devoting attention to the characteristics of strains of *Staphylococcus aureus* and *Staphylococcus epidermidis*,[100] others in the Inoculation Department were investigating more highly pathogenic species of microorganisms, particularly pneumococci and beta-hemolytic streptococci (now designated *Streptococcus pneumoniae* and *Streptococcus pyogenes*, respectively). Infections due to these two groups of organisms were highly prevalent in Great Britain, the United States, and other countries as well. Mortality and morbidity rates were high (see figures 3.1 and 3.2).[101,102]

It is difficult, however, to say with any certainty what the morbidity rates for most infectious diseases were in the 1920s and early 1930s. Records were not kept at all in many areas, and reports from other areas frequently were not submitted to their health departments. The first annual compilation of notifiable diseases in the United States covered the year 1912 and included reports from only nineteen states, the District of Columbia, and Hawaii. In 1916, nine additional states reported, and in

INFECTION DURING CHILDBIRTH AND THE PUERPERIUM

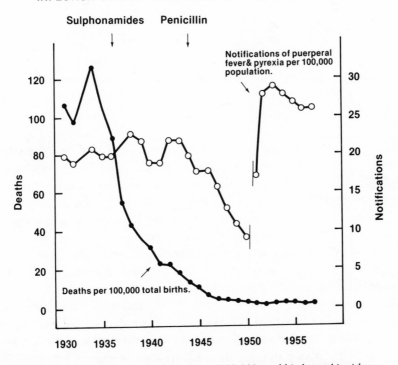

FIGURE 3.1. Puerperal pyrexia—deaths per 100,000 total births and incidence per 100,000 population in England and Wales, 1930–1957. The apparent rise in incidence in 1950 is due to the fact that the definition of puerperal pyrexia was changed in this year. (Reproduced with the editor's permission from M. Barber, Journ. Obstetrics & Gynecology 67 [1960]: 727.)

DEATH RATES
PER 100,000
POPULATION

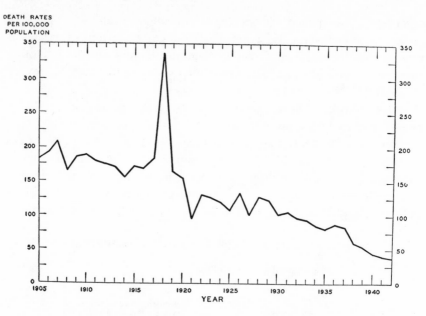

FIGURE 3.2. Trend in pneumonia mortality, New York State, 1905–1942. (From Russell L. Cecil, Textbook of Medicine, *6th ed. [Philadelphia and London: W. B. Saunders and Co., 1943], p. 98.)*

1917, thirty-six states, the District of Columbia, and Puerto Rico submitted statistics. The severe epidemics of poliomyelitis in 1916 and influenza in 1918–19 indicated the need for records, however, and in 1928, for the first time, all states submitted data to the U.S. Public Health Service.[101–103]

Scarlet fever, an infectious disease caused by Group A beta-hemolytic streptococci—highly infectious and contagious—was first reported by all states in 1920, but cases of streptococcal or septic sore throat were not fully reported until as late as 1950.[101,103] The reports of scarlet fever, easily recognized by a characteristic rash that accompanies the infection—a rash due to an erythrogenic toxin produced by the infecting microorganisms— told only part of the story, however. Not all strains of hemolytic strepto- cocci produce the characteristic rash, and it is not produced in all persons infected with toxin-producing strains. The same strains of beta-hemolytic streptococci that produce scarlet fever in some people may cause only a streptococcal sore throat in others. The same strains may also cause pneumonia or erysipelas, which is serious and often fatal, but is not characterized by the scarlatiniform rash. Thus, data on the incidence of scarlet fever indicate only a portion of the total number of streptococcal infections that may have occurred. What statistics we do have indicate that the number of reported cases of streptococcal sore throat and scarlet fever in the United States in 1942 was 135,755 (see figure 3.3).[104]

Best statistical data for the end of the nineteenth and early part of the twentieth centuries are perhaps those recorded by the Health Department of the city of New York—a city with a population of more than 3 million by the year 1900. The New York City Health Department established in 1893 a bacteriology laboratory for the diagnosis of infection,[105,106] and with Herman M. Biggs and William H. Park in charge, the laboratory quickly

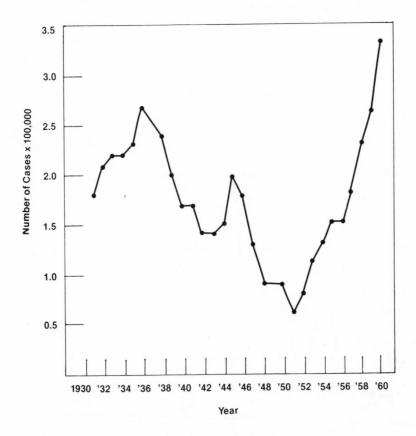

FIGURE 3.3.   Notifiable diseases, United States. Summary of reported cases of streptococcal sore throat and scarlet fever, 1930–1959. (Based on data provided in: Annual Summary—1979: Morbidity and Mortality Weekly Report, September 1980, vol. 28, no. 54, pp. 14–17.)

became an important influence in public health control. Park started making diphtheria antitoxin in horses in 1894. Soon after, he tried to treat a few patients with antitoxin and demonstrated that persons in contact with diphtheritic patients often become carriers of the infecting organism (*Corynebacterium diphtheriae*). In 1907, he confirmed the observations of Theobald Smith of the Rockefeller Institute for Medical Research in New York, who even now is considered America's greatest microbiologist. Smith had shown that injections of diphtheria antitoxin that had been produced in horses, when mixed with minute amounts of diphtheria toxin, would induce in children sufficient antitoxin formation to protect them against the disease (see figures 3.4 and 3.5).

The department's interest simultaneously focused on other infectious diseases as well, and it was during this period that Park instituted methods for recording accurately the prevalence of many infectious diseases. The New York City and U.S. Public Health Service compilations established the then high incidence of pneumococcal, streptococcal, and venereal diseases, against which penicillin later proved so effective.

Also of major importance at the time was puerperal sepsis (childbed fever), another beta-hemolytic streptococcal disease. Highly contagious, it

MENINGOCOCCAL INFECTIONS — Reported Cases per 100,000 Population by Year, United States, 1920–1976

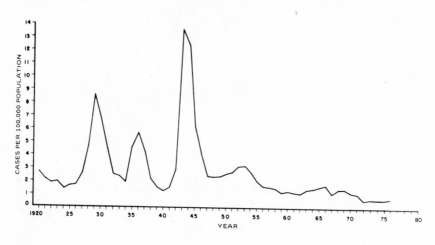

FIGURE 3.4. Meningococcal infections—reported cases per 100,000 population by year, United States, 1920–1976. (From: MMWR Morbidity and Mortality Weekly Report, *Center for Disease Control: Annual Summary 1976 [August 1977, vol. 25, no. 53], p. 50.)*

DIPHTHERIA – Reported Cases and Deaths per 100,000 Population by Year, United States, 1920–1976

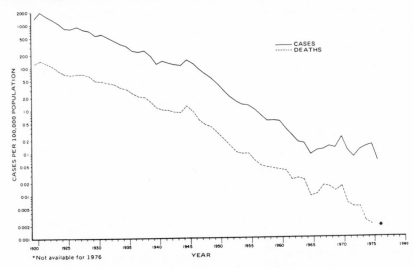

FIGURE 3.5. *Diphtheria—reported cases and deaths per 100,000 population by year, United States, 1920–1976. (From:* MMWR Morbidity and Mortality Weekly Report, Center for Disease Control: Annual Summary 1976 *[August 1977, vol. 25, no. 53].)*

spread rapidly from patient to patient and to others around them. For a number of years, it occupied the attention of Leonard Colebrook in England, while Edgar Todd, also in London, investigated the role of these same infectious agents in the development of rheumatic fever. Long before most others, Colebrook and Todd were aware of the causative relationships between beta-hemolytic streptococci and the diseases they studied. Yet, there is no indication that either of them considered trying penicillin for the treatment of these infections, despite their having worked alongside Fleming.

1939: The Eve of the Penicillin Era

By 1939, more than ten years after Fleming's discovery of penicillin, conditions were ripe for its development. Many of its biological properties were known, though it had not been produced in significant quantity or in any degree of purity. In the case of antimicrobial substances including the sulfonamides, the ability of at least some of them to protect against and cure infections due to susceptible microorganisms had been established. Chemotherapy had become an accepted means of treating infectious disease.

Moreover, the importance of symbiotic relationships between microorganisms in nature had been recognized, and the ability of microorganisms to produce chemicals that could inhibit the growth of other organisms in vitro or in vivo was accepted. The use of antibiotics had been introduced.

Given that infection rates were so high, the need for agents to control the growth and pathogenicity of infecting organisms was great. Thus, the stage was set for the development of penicillin.

It has been said many times that penicillin was developed by the U.S. government and by the United States pharmaceutical industry. This is only partly true. It has also been said that the U.S. government provided the impetus, the drive, and the money. Most important was the money which came at a time when the German bombings of England precluded Britain's development of what unquestionably had been a British discovery and an important British contribution to the war effort.

Actually, the medical needs of the time provided the real impetus. Even without government money or intervention, it was inevitable that penicillin, or a similar drug, would have been developed, once penicillin's capabilities were known.

It is unclear to what extent Fleming understood the interrelationships among his studies on penicillin and others that were conducted: those on prontosil and the sulfonamides; on microbial antagonisms as reported by Waksman, Welsch, Dubos, and others; on the properties of penicillia as reported by Thom; or on the actinomycetes as reported by Waksman.

Fleming, however, was a knowledgeable microbiologist, who was surrounded by other knowledgeable scientists. By 1939, he knew of and surely must have been impressed by the observations of his friend, Leonard Colebrook, who had so clearly demonstrated the efficacy of sulfanilamide and its related compounds in the treatment of puerperal sepsis. By that year, too, the efficacy of chemotherapy as epitomized by the sulfonamides was accepted throughout the world. (It is interesting that although Fleming conducted a few experiments with sulfanilamide, he made no concerted effort to evaluate the various compounds that followed.)[107] Knowing him later, however, it seems clear to me that he must have at least sensed at an early time the importance of his own findings on penicillin.

It must be borne in mind that Fleming was an academician of the 1920s and 1930s; he was a physician interested in infection and its control. At that time he could have had no understanding of the necessarily close relationship between those who observe microbiological phenomena and clinical events and those who "develop drugs." The people working with him in the laboratory, moreover, were mostly young students, the majority of them en route to clinical practice and not yet leaders in their field.

If penicillin had been discovered in France, where Pasteur's studies on fermentation had left such a lasting impression, or in Germany, where the extensive work of Paul Ehrlich and the I. G. Farbenindustrie had evoked strong interest in chemicals with chemotherapeutic properties, the situation might well have been different.

4

Background Events Leading to Major Involvement of the United States in Penicillin Production, 1939–1941

Neither Howard Florey nor Ernst Chain, at the Sir William Dunn School of Pathology in Oxford, recognized in 1938 the far-reaching importance of their decision to undertake a systematic study of microbial antagonisms and antimicrobial agents. Nor did they realize then the importance of their decision to investigate first the properties of penicillin, produced by *Penicillium notatum* and described by Alexander Fleming in 1929. They could easily have started with any number of other substances, ones of little practical importance and of interest only because of the role they play in the metabolism of living microorganisms. But Florey and Chain realized that any substance that could lyse staphylococci, which Fleming had shown penicillin could do, warranted investigation. The similarity between the lytic action of lysozyme and that of penicillin did not escape them.

Certainly Florey and Chain did not recognize the magnitude of the project they were undertaking. Within a few months, what had started as a small research investigation of only academic interest required the full attention of more than ten or a dozen highly trained and experienced investigators, representing several scientific disciplines.

David Wilson commented in *In Search of Penicillin*, published in 1975: "The standard version of the penicillin story is this: penicillin, the first of the antibiotic drugs. First observed by Sir Alexander Fleming in 1928, when he noticed that a stray mold had killed germs on areas of his culture plates. Developed by Lord Florey and Professor Sir Ernst Chain at Oxford in 1940. Mass produced by the U.S. pharmaceutical industry, it saved the lives of thousands of Allied servicemen and came into world use after the end of World War II."[1]

This was the official story for many years, and it sounds so easy as one reads this brief paragraph. But few other scientific projects required the talents, knowledge, and perseverance of so many people as the development of penicillin did. And few, if any, scientific studies required the cooperation of persons from so many scientific disciplines.

The chances of success for the project in the early 1940s were small. Costs were exceedingly high. The Great Depression of the 1930s had not been forgotten, and a war that was to involve all the major countries of the world had broken out. Drug research was just beginning to undergo the change that would result in its becoming a function of universities, medical centers, governmental laboratories, and the pharmaceutical industry rather than of individuals as a by-product of other activities. But with infection so prevalent and mortality rates so high, the need for methods of treatment and for control of infection was great. The studies of the sulfonamides had shown that some serious infections could be controlled if proper drugs were available.

At Oxford, 1939 to 1941

Howard Florey had moved to Oxford in 1935 to become the Sir William Dunn Professor of Pathology, succeeding Professor George Dreyer who had died in 1934. Once Florey was at Oxford, he started immediately to build a team for the study of lysozyme. He was convinced that lysozyme warranted investigation, but its antibacterial properties per se were of little interest to him. He wanted to elucidate the function of the lymphatic system.

Florey had trained as a physiologist, and he brought to his new department an interest in experimental pathology. Aware of the need for interdisciplinary approaches to studies in pathology, he wanted a biochemist who could work independently, yet collaborate with him, to make his studies of lysozyme successful. He knew that lysozyme had to be purified if its mode of action was to be elucidated and its role in nature understood.

Developing an interdisciplinary staff of skilled investigators and obtaining money to support them was Florey's first concern after moving to Oxford. Financial support came in small amounts from the Nuffield Trust, the Medical Research Council, and the Rockefeller Foundation.

Florey was fortunate in that, soon after moving to Oxford, he met Ernst Boris Chain, a student of Sir Frederick Gowland Hopkins who was head of the Sir William Dunn School of Biochemistry at Cambridge.

59

Chain—born in Berlin in 1906, his mother German, his father a Russian industrialist and chemist—had graduated from the Friedrich-Wilhelm University where he trained in chemistry and physiology. Simultaneously, he trained for a career as a concert pianist. Perhaps because of the political and economic situation in the early 1930s, he chose to pursue a career in science, but retained his interest in music throughout his life.[2,3]

Chain was forced to leave Nazi Germany in 1933. He emigrated to England and affiliated first with the University College Hospital in London and then, after a short time, with Sir Frederick's department at Cambridge. By 1935, he had completed two doctoral theses and was seeking a permanent position, just when Florey began looking for an associate to develop the biochemical section of the Department of Pathology at Oxford. Sir Frederick highly recommended Chain, who soon moved to Oxford. Later, Chain would write:

I joined Florey's staff in the middle of 1935, a few months after he had been appointed Head of the Sir William Dunn School of Pathology, in succession to the Danish pathologist, Professor Dreyer who had died shortly after the construction of the new pathology building was completed, a very nice somewhat continental, spacious and (for the times) well-equipped building. The task I was given by Florey was to organize a biochemical section in the department. Though he was not a biochemist himself (he was more a physiologist than a classical pathologist by training, very much influenced by his teacher, Sir Charles Sherrington for whom he had great respect), he was firmly convinced that biochemistry had a very important role to play in the progress of functional pathology.... The history of research... in this subject...has proved him right, hence his decision to create a biochemical section in his department right at the beginning of his activity, as head of the Department of Pathology at Oxford, undoubtedly was a far-sighted move.[2]

At Oxford, Chain first worked on the biochemical effects of some snake venoms. Later, having decided to extend his research to studies of the metabolism of cancer tissue, he recognized the need for assistance with microanalyses and suggested to Florey that Dr. Norman G. Heatley, then completing his scientific training at Cambridge, be invited to join the department. Heatley became a member of the team on October 2, 1936.[4] Dr. A. D. Gardner and Dr. Jean Orr-Ewing, who had worked in the department under Professor Dreyer, remained there after Florey succeeded Dreyer as professor of pathology. Dr. E. A. H. Roberts had moved in 1935 from the Dyson Perrins Laboratory of Organic Chemistry, headed by Sir Robert Robinson, and Dr. A. G. Sanders came into the

PLATE 4.1.   Prof. Norman G. Heatley, M.A., Ph.D. (Photograph by Maggie Bristol, ARPS.)

PLATE 4.2.   Lady Mary Ethel Florey, M.B., B.S. (Courtesy of Paquita Flora McMichael.)

PLATE 4.3.   Prof. Sir Edward P. Abraham, M.A., D.Phil. (Courtesy of the American Society for Microbiology.)

department in 1936 to help in research in experimental pathology. The Honorable Dr. Margaret Jennings, daughter of Lord Cottesloe and wife of Dr. D. A. Jennings, a gastroenterologist in London, moved to Oxford when her husband transferred his affiliation from the Royal Free Hospital in London to the Radcliffe Infirmary in Oxford. She joined the group at the Sir William Dunn School of Pathology in 1936 and remained in the department for many years. In June 1967, she became Lady Margaret Florey, after Lady Mary Ethel Florey, who conducted the earliest clinical studies with penicillin, died on October 10, 1966.

Thus, by late 1936, Florey had brought together most of the scientists who, as a team, were destined to acquire world renown for their early and classical studies of penicillin. Only Dr. E. P. Abraham, a chemist, and Dr. Charles Fletcher, a clinician, joined the team later in 1940.

Strangely, penicillin had first been produced at the Sir William Dunn School in 1930—five years before Florey moved to Oxford. George Dreyer, Florey's predecessor, had been interested in bacteriophages— viruses with specific affinity for bacteria that are found in association with essentially all groups of bacteria and blue-green "algae." In 1929, he had

THE LANCET] PROF. FLOREY AND OTHERS : PENICILLIN AS A CHEMOTHERAPEUTIC AGENT [AUG. 24, 1940 227

RESULTS OF THERAPEUTIC TESTS ON MICE INFECTED WITH *Strep. pyogenes, Staph. aureus* AND *Cl. septique*

Expt.	—	Dose of infecting culture (c.cm.)	Interval before starting treatment (hrs.)	Duration of treatment	Single dose (mg.)	Total dose (mg.)	No. of mice	6	12	24	2	3	4	5	6	7	8	9	10
								hours			days								
				Strep. pyogenes—Lancefield, Gp. A.															
1	Controls	0·5	..	12 hrs.	2	10·0	25	..	15	9	8	6		5		4			4¹
	Treated	0·5	1			50		..	..		49	42		34	30	28		26	25
2	Controls	0·5²	..	45 hrs.	0·5	7·5	25	24	3	0³	..	..	..	..	..	..	..	..	0
	Treated	0·5	2			25	24	..	..	..	..	..	..	..	..	..	..	24
				*Staph. aureus*⁴															
1	Controls	1·0	..	55 hrs.	0·5	9·0	24	21	1	0³	..	11	..	10	..	..	..	..	0
	Treated	1·0	1			25													8
2	Controls	0·2²	..	4 days	0·5	11·5	24	23	15	5	0	..	21	..	..	..	..	..	0
	Treated	0·2	1			24													21
				Cl. septique															
1	Controls	see text	..	..	..	..	25	..	21	0⁵	..	..	..	..	..	..	..	..	0
	Treated	..	1	10 days	0·5	19	25	..	..	24	21	..	18	..	..	..	..	..	18
		..	1	10 ,,	1·0	38	25	..	..	..	24	..	..	24	..	..	..	..	24

1. A control mouse which was killed by mistake at 24 hrs. is counted as a survivor. Heart-blood culture strongly positive.
2. Between experiments 1 and 2 the virulence of the organism was raised by passage.
3. Controls all dead within 16 hrs.
4. A bovine strain exceptionally virulent to mice kindly supplied by Dr. H. J. Parish of the Wellcome Laboratories.
5. Controls all dead within 17 hrs.

PLATE 4.4.   First experiment showing chemotherapeutic effects of penicillin. (Reproduced with permission from: E. Chain, H. F. Florey, et al., "Penicillin as a Chemotherapeutic Agent," Lancet 2 [August 24, 1940]: 226–28.)

seen Fleming's paper on penicillin and had obtained a reprint of the article and a culture of mold. He soon showed that Fleming's strain of *Penicillium notatum* was not a carrier of bacteriophage, but he continued to grow the mold and to use it to obtain plaques of a standard size for quantitation of other bacteriophages. Miss Campbell-Renton, who was associated with Dreyer for many years, continued this work after his death. Fleming's contaminant thus was cultured in the department at Oxford from 1930 on, and Chain's first attempts to produce penicillin were made with a sub-culture of this strain.[2]

The first scientific report on penicillin—by Ernst Chain, Howard Florey, A. D. Gardner, N. G. Heatley, M. A. Jennings, J. Orr-Ewing, and A. G. Sanders—was published in the August 24, 1940, issue of *Lancet*, a highly regarded British scientific and medical journal.[5] In it they reported the methods they devised to obtain significant amounts of penicillin and to rapidly assay its growth inhibitory power. They also reported initial tests of its toxicity in three animal species (mice, rats, and cats) and preliminary studies of its chemotherapeutic action in mice experimentally infected with

PLATE 4.5.   The Hon. Margaret A. Jennings, M.A., M.B., D.Phil., who became Lady Margaret Florey in June 1967. With Dr. William M. M. Kirby of Seattle, Washington, at the 11th International Congress of Chemotherapy and the 19th Interscience Conference on Antimicrobial Agents and Chemotherapy, Boston, Massachusetts, 1979. (Courtesy of the American Society for Microbiology.)

strains of beta-hemolytic streptococci, *Staphylococcus aureus*, and *Clostridium septique* (an anaerobe often found in infected wounds). The number of animals used in their experiments was small, but the results were striking:

The results are clear-cut, and show that penicillin is active in vivo against at least three of the organisms inhibited (by penicillin) in vitro. It would seem a reasonable hope that all organisms inhibited in high dilution in vitro will be found to be dealt with in vivo. Penicillin does not appear to be related to any chemotherapeutic substance at present in use and is particularly remarkable for its activity against the anaerobic organisms associated with gas gangrene.[5]

The experiment on which this conclusion was based was started on July 11, 1940. The decision to proceed with in vivo studies with penicillin was made at an earlier date, however. On May 25, 1940, eight mice were infected intraperitoneally with hemolytic streptococci and were then divided into three groups.[6,7] One group (A) consisted of two mice each of which received subcutaneously five doses of 0.1 ml each of a solution containing 50 mg of penicillin powder per milliliter at 1, 3¼, 5¼, 7⅓, and 11 hours after infection. The second group (B) also consisted of two mice, each of which received one dose of 0.2 ml of the same solution 1 hour after infection. Four mice in a third group (C) were untreated and served as controls. The penicillin extract used probably contained between 2 and 4 units per milligram. In group A, one animal survived thirteen days, and one survived indefinitely. In group B, one survived for two days and one for six days. In group C, all died within 16½ hours. On May 27, and again on May 28, other experiments involving small numbers of animals were initiated, always with striking results. It was on the basis of the May 25 experiment, however, that the decision was made that studies should be directed quickly toward evaluation of the efficacy of the substance as a chemotherapeutic agent in humans.

Thus, Florey, Chain, and their associates at a very early stage made the transition from a basic in vitro study of what they thought was an enzyme and its antibacterial effects to a practical study of the chemotherapeutic action of the substance. Fortunately, Chain was wrong in assuming that penicillin was an enzyme, for, as he remarked some years later, had it been an enzyme, it probably would have been a protein capable of producing severe hypersensitivity reactions in animals and man.[2,4]

In the August 6, 1941, issue of *Lancet*, the Oxford group described in detail conditions required for the effective production of large amounts of penicillin by their strain of *Penicillium notatum* and for assay of material they considered suitable for therapeutic use in man.[8] In addition, they reported preliminary observations on the absorption and excretion of the drug in rabbits, cats, and man, and showed that it is excreted in high concentration in the urine of all three species. They noted a high concentration, moreover, in cats' bile, but less in that of rabbits. They also presented evidence of its low toxicity when applied directly to body tissues. No toxic effects attributable to penicillin were observed when the substance was injected intravenously. The crude penicillin products contained a pyrogenic substance, however, which they found could easily be removed.

In the experience of the Oxford investigators, growth in vitro of many pathogenic bacteria was prevented by their partially purified penicillin when it was diluted one in 1 million or more. Other organisms showed lower degrees of susceptibility and still others were quite resistant. Proof was given of the inability of blood, pus, and tissue derivatives to prevent the action of penicillin. Adaptation of strains of *Staphylococcus aureus* to growth in high concentrations of the substance was reported, and this was shown not to be due to the development of a penicillin-destroying enzyme. Data were presented on a comparison of the antibacterial activity of penicillin and of the sulfonamides, and reasons were offered to explain why penicillin could be expected to operate when the sulfonamides were ineffective.

Of greater importance, though, were Florey's comments concerning the therapeutic use of penicillin:

During the course of some therapeutic trials in human infections it has proved possible to secure and maintain a bacteriostatic concentration of penicillin in the blood without causing any toxic symptoms. After intravenous administration a large proportion of the active substance can be recovered from the urine and used again.

Penicillin was given intravenously to five patients with staphylococcal and streptococcal infections and by mouth to one baby with a persistent staphylococcal urinary infection. It was also applied locally to four cases of eye infection. In all these cases a favourable therapeutic response was obtained.[8]

This second report by Florey, Chain, and their associates covered many aspects of the chemotherapeutic action of penicillin. The work was conducted by a large number of collaborating investigators, and Florey gave full credit to each person individually for his or her contributions to the study.

The work was planned and started by E. Chain and H. W. Florey. The chemical and biochemical part of the work was carried out in the main by E. Chain and E. P. Abraham. N. G. Heatley devised the assay method and developed and supervised the production of penicillin. M. A. Jennings and H. W. Florey have conducted the biological tests except those especially mentioned in the text. A. D. Gardner has conducted the bacteriological investigations and made some special observations on the growth of the mould. C. M. Fletcher (of the Nuffield Department of Medicine) has been in charge of the administration to man. The successful conduct of the work to its present stage has only been made possible by the closest collaboration of all concerned. We wish to thank the physicians and surgeons who placed their cases at our disposal. . . . We wish also to acknowledge the work of the following technicians, without whose efforts adequate supplies of penicillin could not have been produced: D. Callow, R. Callow, M. Lancaster, P. McKegney, E. Vincent. . . . We are indebted [too] to the Medical Research Council for a grant towards the expenses of large-scale production, and to the Rockefeller Foundation for providing for technical assistance and expenses.[8]

Raymond B. Fosdick, for many years president of the Rockefeller Foundation, later remarked in his history of the foundation:

A small grant to Oxford University in 1936. . .was given in response to the application of a professor of pathology who explained that he had recently engaged a German refugee biochemist to collaborate with him on problems of "chemical pathology" and needed 250 British pounds with which to purchase laboratory equipment. The professor was Howard W. Florey, the refugee was Ernst Chain, and the eventual outcome of their research was the purification of penicillin and the proof of its clinical value. Oxford University furnished by far the greater part of the support—the Rockefeller Foundation in its succession of grants was a very minor partner—but. . .a proposal which the Natural Sciences Division accepted as an opportunity in pure biochemistry. . .produced a result promptly, importantly, and directly useful to medicine. . . .

Utilitarianism. . .was never the yardstick used by the Rockefeller Foundation in making its grants. Projects were not espoused or rejected according to the degree with which they promised an immediate practical result. The test was knowledge: Is the proposed study likely to add to man's understanding of the living world of which he is a part? As light increases, the darkness recedes to greater distances; similarly, the increase of fundamental knowledge spreads in all directions and carries the power to illumine many problems. That was the faith in which, in the early 1930s, the Foundation entered upon its plan to advance experimental pathology.[9]

Ernst Chain and Howard Florey profited greatly by this thinking and by the philosophy of the Rockefeller Foundation's trustees and staff. The world profited, too.

On January 27, 1941, Florey, Chain, and their associates administered their first dose of penicillin parenterally to a human. On that day, Dr. C. M. Fletcher administered 100 mg of material "of which 10 mg had had no ill effect on a mouse weighing 20 grams" to a patient in the Radcliffe Infirmary. There were no untoward effects other than a febrile response to the pyrogen-containing crude preparation used. The patient was not suffering from any infection, and the injection was administered only to evaluate the possible toxicity of the preparation (see table 4.1).[6]

In the United States, 1939 to 1941

Prior to September 1940, Martin Henry Dawson, Karl Meyer, and I, at the College of Physicians and Surgeons in New York, had been investigating for some time the properties of pathogenic strains of hemolytic streptococci and the diseases they produce. Immediately prior to and during 1940, our interest had centered largely on the powerful enzyme—hyaluronidase—produced by some (not all) strains, on its possible role in the pathogenesis of streptococcal diseases, and on its ability to hydrolyze hyaluronic acid, a mucopolysaccharide component of vitreous humor. Hyaluronic acid is present also in joint and certain other body fluids and in some tissues, including particularly connective tissue. It is also present in the capsule (outer component) of virulent Group A hemolytic streptococci. Hyaluronidase had been under investigation, also, by Ernst Chain in Oxford—though for other reasons.

Henry Dawson was a clinician with special interest in microbiology, infectious disease, and the rheumatic diseases, particularly rheumatoid arthritis which some thought might be due to hemolytic streptococci. Dawson had become especially interested in subacute bacterial endocarditis, an infection of the heart valves occurring often in patients with rheumatic fever, usually (although not invariably) caused by streptococci and (at the time) always fatal. He recognized immediately upon reading the first publication by Florey, Chain, and their associates that if penicillin could be produced in adequate amounts, it might be effective in the treatment of streptococcal and pneumococcal infections in general and subacute bacterial endocarditis in particular.[10,11]

Within two weeks after the publication of the first report reached the United States, Dawson made contact with Ernst Chain to obtain a subculture of his penicillin-producing strain of *Penicillium notatum*. At the same time, he obtained cultures from Roger Reid who had worked on penicillin in the United States in the early 1930s. On September 23, 1940,

TABLE 4.1   Penicillin Highlights: At Oxford

At Oxford	Interest in Penicillin Grows
1938 to 1939 Literature survey by Ernst Chain stirs interest in penicillin: Howard Florey and Ernst Chain start investigation of penicillin as a microbial antagonist in nature.	**February 1941** First clinical trial (in six patients) initiated by Florey: staphylococcal and streptococcal infections that were uncontrolled by sulfonamides or surgery.
September 1939 Great Britain declares war with Germany: World War II begins. Penicillin study takes on new meaning, new goals.	**February to June 1941** Norman Heatley and associates at Oxford develop methods for greater production of penicillin: Florey and associates complete first clinical trial.
May 1940 First experiments performed in mice show chemotherapeutic potential.	**May 1941** M. H. Dawson and associates (Columbia University, New York) report to Society for Clinical Investigation on therapeutic potential of penicillin.
August 1940 Oxford team publishes methods for production of penicillin and evidence of its chemotherapeutic action.	
October 1940 Oxford results confirmed by Dawson and associates in United States. First patient injected with penicillin at Columbia Presbyterian Hospital, New York.	**June 1941** Imperial Chemical Industries interested. Florey fails to respond to company's offer of manufacturing assistance. Florey and Heatley depart for United States to seek manufacturing assistance.
January 1941 Dr. C. M. Fletcher administers first dose of penicillin to a patient in Oxford.	**July 1941** The United States responds. Penicillin research begins at U.S. Department of Agriculture's Northern Regional Research Laboratory.

(*Continues*)

...dent Is Set

...1941

On ...
Medica...
collaborati...
8th, represen...
selected pharmac...
Company, E. R. Squi...
Pfizer & Co., and Lederle
Laboratories) hold first meet...
penicillin production. Collabora...
within the industry was needed.

December 1941
NRRL reports improved penicillin
fermentation techniques, increased
production, probable feasibility of
submerged fermentation.

June 28, 1941
In the United States, the War
Production Board had been established,
with rubber at top of priority list.
International collaborative research on
penicillin no longer feasible.

...United States enters
...r imports
...
...han 25
...er-tired
...lar,

...epartment
...l, rubber,
and petro... ...exchange
information freely. Co...te interest
in proprietary rights to discoveries was
abandoned and information flowed
freely. Necessary supplies became
available. A precedent was set for
exchange of information among
competing industrial firms—a precedent
also important to those attempting
penicillin production.

we initiated our first experiments on penicillin, using the culture received from Roger Reid. We did not plan then to study fermentation processes for the production of penicillin. Rather, we naively undertook "to make some penicillin." I handled the microbiological aspects of the study, Karl Meyer served as chemist, and Henry Dawson was coordinator and clinician.

Less than one month later—on October 15, 1940—first doses of penicillin were administered parenterally to two patients, Aaron Alston and Charles Aronson, at the Presbyterian Hospital in New York. The material used was a crude, slightly purified (concentrated) preparation in butyl alcohol. Initially 0.1 ml was given intracutaneously to each patient. A few hours later, each received 1.0 ml subcutaneously, and on the following day, each received 5.0 ml, again subcutaneously and with local "soreness" only. On December 31, 1940, Alston received intramuscularly 1.0 ml of another preparation of penicillin in propylene glycol. Beginning January 11, 1941, he received daily intravenous injections totaling 4.25 ml over a six-day period.

Simultaneously, penicillin was administered intravenously to another patient at the Presbyterian Hospital, a Mr. Conant. Initially the drug was given once daily and then twice daily, beginning on January 18 and continuing through January 26, the total amount coming to 36 ml over the sixteen-day period.[10]

The penicillin preparations used in our early trials were prepared by Dr. Karl Meyer. All were heavily pigmented and extremely crude, but highly active. Dilutions of 1:50,000 to 1:200,000 caused complete inhibition of growth of a standard strain of Group A beta-hemolytic streptococci. All were effective also in the treatment of animals experimentally infected with this strain.

Dawson was the first to employ penicillin parenterally with a therapeutic response as an immediate goal. Although the amount administered to the first patients was too small to achieve a therapeutic response, he believed that in patients with bacterial endocarditis even a decrease in the number of infecting organisms per milliliter of blood would be significant.[11–14]

At first, only the low toxicity of the crude and impure penicillin preparations used was noted. It was not until March 26, 1942, that Dawson observed a clear-cut therapeutic response in a patient with endocarditis. An initial dose of 30 mg followed by doses of 20 mg once every three hours was administered intravenously. The penicillin employed was Pfizer's lot no. 792 (sodium salt), active in dilutions of 1:10,000,000 and 1:40,000,000 when tested in vitro against strains of hemolytic streptococci and staphylococci, respectively. The number of infecting organisms per milliliter of the patient's blood fell from 650 just prior to start of treatment to 2 less than twenty-four hours later.

From January 1941 on, it had been clear that if properly purified and available in sufficient quantity, penicillin could be used parenterally and probably effectively in the treatment of infections due to susceptible microorganisms. The small amounts of penicillin made by Karl Meyer, Henry Dawson, and myself during 1941 initiated the early clinical trials of penicillin in the United States.

The first scientific paper from our laboratories on penicillin was presented on May 5, 1941, at the annual meeting of the American Society for Clinical Investigation (popularly known as the "Young Turks"), a forum of great importance, for all papers were critically reviewed before acceptance for presentation. Data were presented that confirmed the in vitro and in vivo activity of the substance, and it was also reported that penicillin had been administered to four patients with subacute bacterial endocarditis and to eight patients with chronic staphylococcal blepharitis.[14,15] The report said:

Sufficient material was not available for adequate therapy in the cases of subacute bacterial endocarditis. However, no serious toxic effects were observed. The results in local eye infections were most satisfactory. One patient who proved resistant to sulfathiazole showed a prompt response. Further clinical trials are under way...it would appear that penicillin is a chemotherapeutic agent of great potential significance. It is effective *in vivo* as well as *in vitro* on both aerobes and anaerobes in remarkably small concentrations. It is apparently not inhibited by blood and serum, nor by pus and other substances which are known to inhibit the sulfonamides. It appears to possess little toxicity. Penicillin probably represents a new class of chemotherapeutic agents which may prove as useful or even more useful than the sulfonamides.[15]

Dawson's paper at Atlantic City on the action of penicillin was reported in detail by Steven M. Spencer in the *Philadelphia Evening Bulletin*[16] and by William L. Lawrence in the *New York Times*.[17] From then on, interest in the drug was enormous. Dr. Wallace Herrell who had been present to hear the report returned to the Mayo Clinic to immediately start work on the drug.[18] Others reacted also. Jasper Kane and Gordon Cragwall of Chas. Pfizer & Co. (now Pfizer, Inc.), then a small manufac-

THE EVENING

GERM KILLER FOUND IN COMMON MOLD

Physician says Penicillin may Prove more Useful than Sulfanilamides

By STEVEN M. SPENCER
(Of The Bulletin Staff)

Atlantic City, May 5.—From a common mold which grows on stale bread and which gives Roquefort cheese its highly prized green color scientists have prepared a germ-chasing medicine which "may prove as useful or even more useful than the sulfanilamide drugs."

This appraisal of penicillin, extracted from the mold known as Penicillium, was made today by Dr. Martin H. Dawson, associate professor of medicine in the College of Physicians and Surgeons, New York, speaking at the 33d annual meeting of the society for clinical investigation, at Haddon Hall.

Dr. Dawson reported experiments and treatments conducted with the mold medicine by himself and his associates, Drs. Gladys L. Hobby, Karl Meyer and Eleanor Chaffee.

Penicillin was first prepared and was named by Dr. Fleming in England, who noticed that a colony of staphylococcus germs accidentally contaminated with spores of penicillium did not grow. This was in 1929 and not much work was done with the material until last year, although now many Englishmen are chasing "strep" and other germs in the nasal passages by using the dried penicillin as a snuff, Dr. Dawson said.

The material, as tested on animals and in a few human cases, appears to produce no toxic effects when given in doses far beyond those necessary to clear up infections, Dr. Dawson reported. This would be a distinct advantage over the sulfanilamide drugs, which are toxic to some people. And the mold medicine is effective against all the germs of the Gram-positive classification, which includes streptococcus, staphylococcus, meningococcus, the pneumonia germs, and many others. It is slightly effective against the Gram-negative bacteria. Dr. Dawson believes it has "unlimited possibilities."

Another medicine obtained from a microscopic form of life, gramicidin from a bacillus found in the soil, has been highly effective in the treatment of ulcerating sores of the leg and other parts of the body as well as in bladder infections and empyema (pus in the chest cavity), it was reported today.

The soil bacillus preparation, isolated two years ago by Dr. Rene Dubos, of the Rockefeller Institute for Medical Research, must not be given by vein, however, Dr. Wallace E. Harrell, of the Mayo Clinic, warned, as it causes breakdown of the red blood cells. Dr. Harrell and his associate, Dr. Dorothy Heilman, concluded it was safe for local application, however, as it is not toxic to the flesh itself.

The New York Times

TUESDAY, MAY 6, 1941. 23 L +

'GIANT' GERMICIDE YIELDED BY MOLD

New Non-Toxic Drug Said to Be the Most Powerful Germ Killer Ever Discovered

TRIED ON HUMAN BEINGS

Physicians Also Told That Safe Way Has Been Found to Use Gramicidin

By WILLIAM L. LAURENCE
Special to THE NEW YORK TIMES.

ATLANTIC CITY, N. J., May 5—A new chemical substance elaborated by a special strain of mold in bread and Roquefort cheese that has proved itself in tests on animals and in preliminary clinical trials on human beings as the most powerful non-toxic germ-killer so far discovered, thousands of times more potent than any of the drugs of the sulfanilamide family, was described here today.

Hundreds of leading physicians from the United States and Canada, attending the annual meeting of the American Society for Clinical Investigation, heard Dr. Martin H. Dawson, Associate Professor of Medicine at the College of Physicians and Surgeons, Columbia University, New York City, report on the new germ-killer.

Associated with Professor Dawson in this work, which physicians here hailed as opening a new chapter in the fight of medical science against bacterial infections caused by the vast host of deadly micro-organisms known as gram-positive bacteria, were Drs. Gladys L. Hobby, Karl Meyer and Eleanor Chaffee.

PLATE 4.6. "Germ Killer Found in Common Mold." (Reproduced from the Philadelphia Evening Bulletin, *May 5, 1941.)*

PLATE 4.7. "'Giant' Germicide Yielded by Mold." (Reproduced from the New York Times, *May 6, 1941, p. 26.)*

turer of chemicals in Brooklyn, N.Y., also heard the report, and soon their company initiated its own studies of penicillin. By the fall of 1941, they were shipping large carboys of penicillin fermentation liquor to us each morning for assay. These significantly increased the amount available to Karl Meyer for extraction and analysis.[19,20]

Our own facilities for producing the substance were little better than Fleming's. Within a few weeks after starting work on penicillin, it was clear that large volumes of fermentation liquor would be needed if sufficient drug was to be available for clinical use. Soon hundreds of two-liter flasks with *Penicillium notatum* growing on a modified Czapek-Dox medium lined every classroom laboratory bench at the Columbia University Medical School. We had no adequately large incubators and no space in our own small laboratory for such large numbers of flasks, but moved in and out of classrooms as the students moved out and in.

Eventually, it was discovered that the flasks could be incubated under the seats of a two-story amphitheater, and at last the penicillium cultures could be grown under stationary conditions—at least during the eight to nine months of the year when room temperatures were within the range suitable for growth of the mold. Karl Meyer and Eleanor Chaffee had set up their still for evaporation of solvents during the extraction of penicillin on the fire escape of the university building. The fumes were thereby vented into the open air of Fort Washington Avenue. Initially, we followed in the footsteps of Fleming and of Florey and his associates, using methods developed by them and adapting the procedures to our needs. Separating active penicillin from the fermentation liquors was not an easy task, as Harold Raistrick and Roger Reid had noted earlier and many others were to learn in the next few years. Later, the techniques we used seemed very primitive, but they provided enough concentrated and partially purified penicillin to convince us of the efficacy of the drug and enough even to help save a few patients' lives.

I do not recall that it ever occurred to us at the time that it might not be possible to find ways of obtaining enough penicillin for treatment of streptococcal infections. Our concern was to establish that it would provide a therapeutic response. All we had going for us in this endeavor was an enthusiastic team consisting of an extremely able chemist, a microbiologist, and a clinician with drive, purpose, dedication, and vision.

Henry Dawson had the foresight to push experimental studies on penicillin, the courage to test the drug in humans at an early stage, and the wisdom to stir the interest of others better able to develop it. Unfortu-

75

PLATE 4.8.   *The author, Gladys Hobby.*

nately, it was his fate never to know of the full impact of penicillin, and the antibiotics that followed, on clinical medicine and on the world in which we live. Martin Henry Dawson died on April 27, 1945, the victim of myasthenia gravis, a progressive disabling disease that had first become apparent early in 1941.

Only the reports of Florey and his team at Oxford and that of Henry Dawson to the American Society for Clinical Investigation (published then only in abstract form) attracted attention during 1940 and 1941. A few other reports appeared in the scientific literature, however. Professor A. D. Gardner[21] described the morphological effects of subinhibitory concentrations of penicillin on rod-shaped organisms. He showed that cultures of *Clostridium welchii* (and some other rod-shaped organisms), when in a fluid medium containing penicillin in concentrations too low to inhibit growth, grew in flocculent form, rather than in a uniformly turbid suspension. Flocculent growth consists of microorganisms growing in tufts, similar to tufts of cotton, suspended in the fluid medium. This type of growth generally is characteristic of organisms with a low degree of pathogenicity. Gardner showed that microscopic examination of the flocculent growth revealed elongated cells that took the form of unsegmented filaments ten or more times longer than average normal cells. Staphylococci and streptococci, on the other hand, when subjected to such subinhibitory concentrations of the substance, showed spherical enlargement of the cells and imperfect fission. The significance of these observations was not recognized for many years, and, indeed, the effect of subinhibitory concentrations of penicillin on bacteria is still under study— more than forty years later.[22] The mechanisms by which these morphological changes occur and the significance of such changes still are not clearly understood.

At about the same time, M. E. Delafield, E. Straker, and W. W. C. Topley (the foremost microbiologist in England at the time and for many years after) made a comparison of sulfathiazole, proflavine, and penicillin in antibiotic snuffs.[23] The purpose of their study was to evaluate the extent to which each of these antimicrobials could control the bacterial flora of the nose and nasopharynx. The snuff contained by weight 1 part penicillin, 5 parts menthol, and 94 parts lycopodium. A crude preparation of the sodium salt of penicillin containing 10 Oxford units per milligram was used. Six applications daily provided a daily dose of penicillin estimated to be about 25 to 35 units. (One unit in 1941 was defined as that amount of penicillin which when dissolved in 1 cc of water gave the same inhibition as an arbitrary standard set up by Florey and his associates in Oxford.)

The four patients treated with penicillin were all heavy nasal carriers of staphylococci, the number of which was greatly reduced during treatment. In three patients, the period of treatment was short (seven to ten days) and the staphylococci returned after the administration of penicillin was discontinued. In the fourth patient, the staphylococci were "almost abolished" for eight days after the short course of penicillin. As the authors mentioned, though, one could not be sure that there was not a carryover of penicillin on the swabs used for culture collection. But, although these patients received topical therapy only, the effects observed were striking, particularly in view of the small amount of substance administered. Three of the patients, however, may have been the first to show persistence of viable microorganisms after apparently adequate penicillin administration.

Finally, in December 1941, both Ernst Chain[24] and Edward Abraham[25] reported preliminary data on the mode of action of penicillin. With Henry Dawson and Karl Meyer, I reported during the same month to the Society of American Bacteriologists (now the American Society for Microbiology), at a meeting in Baltimore, that preparations of penicillin made by us, using a modified Czapek-Dox medium containing dark brown sugar (in lieu of cane sugar), produced marked antibacterial effects against a wide variety of microorganisms:

A concentration of 0.01 to 0.1 microgram per milliliter is sufficient to inhibit the growth of 2,500,000 hemolytic streptococci (group A). In certain instances, the effect of penicillin appears to be bactericidal while in other cases it appears to be bacteriostatic. The results of experiments on its mode of action...[were] described and the *in vitro* effect of penicillin compared with that of the sulfonamides and gramicidin.[26]

These data were published in abstract form during 1941 and in greater detail in June 1942, when evidence was presented to show the spectrum of activity of penicillin in vitro and in vivo and to support the statement made earlier (1941) that penicillin is effective only against actively multiplying microorganisms.[27-29] At the time, it seemed probable, too, that the primary action of penicillin was to prevent cell wall synthesis (see chapter 1).[30] Not until 1942 did Karl Meyer describe the extraction procedures used in these studies:

The culture medium is adjusted to pH 3–4, saturated with ammonium sulfate and extracted with chloroform. The active agent is removed from the concentrated chloroform extract by phosphate buffer at pH 7.2. Extraction with chloroform and buffer is repeated and the less acidic pigments separated from the most active fraction by chloroform extraction at different acidities. Penicillin is obtained from the concentrated extracts either as the free acid by precipitation from petroleum ether, or as the ammonium salt by saturation of a chloroform-benzol solution with dry ammonia gas. If the precipitation of the free acid is slow, it separates in yellow thick whetstone shaped crystals. The ammonium salt forms a dark yellow micro-crystalline powder. In solution, penicillin, especially the free acid, is rather rapidly inactivated. The ammonium salt is more stable. In dry form both acid and salt keep only *in vacuo*. This procedure has given uniform results and a yield of over 50 per cent. of the original potency.

A considerable increase in stability of the solutions was obtained by acetyla-tion or benzoylation of the ammonium salt. The free acids of the acyl derivatives form fine needles which have about the same *in vitro* activity as the mother substances. . . .

Biologically our preparations are inactive against *E. coli*. The minimal concentration showing activity against 2 to 3 million hemolytic streptococci per cc is at a dilution of 1 : 32 million. This corresponds to about 240 Oxford units per mg. The Oxford standard has an activity of 42 units per mg.[31]

Thus, the observations of Fleming, made ten to twelve years earlier, and those of Florey, Chain, and their associates were confirmed in the United States. The work of the Columbia University team had aroused the interest of Chas. Pfizer & Co., Inc., as well as that of others. Selman Waksman, moreover, with his long interest in microbial antagonisms and antimicrobial substances, had stirred the interest of scientists at Merck & Co., Inc. and Dr. Geoffrey Rake at nearby E. R. Squibb & Sons had initiated some experiments with the penicillin-producing organism, *Penicil-lium notatum*. Slowly studies were starting up that were to lead ultimately to mass production of the drug.

These events occurred in 1940 and early 1941, only because of the promise penicillin offered. There had been no government intervention to this date and no wartime pressures. All that remained to be done was to demonstrate on a wide basis the clinical usefulness of penicillin and to develop methods for its large-scale production. The magnitude of these remaining aspects became apparent as time passed.

It now seems to me quite possible that the impetus provided entirely by the climate of the medical community, and the morbidity and mortality rates in the United States at the time, would have led sooner or later to the successful commercial development of penicillin for clinical use. Unques-tionably, developments would have come more slowly than they did, but even among the civilian population, the need for an effective chemotherapeutic agent was great. The time was ripe, and penicillin offered promise.

The pace of research on the drug suddenly quickened during the last half of 1941, when Sir Howard Florey (later Lord Florey) came to the United States with his associate, Dr. Norman Heatley, "to explain to American scientists the experience of the Oxford Laboratory in the production of penicillin"[32,33] and to urge the U.S. pharmaceutical companies to enter into mass production. The pace further quickened when the United States was plunged into World War II by the Japanese attack on Pearl Harbor and the Philippine Islands on December 7. The development of penicillin thereafter became by every measure the greatest achievement of the U.S. pharmaceutical industry and probably of the medical profession during World War II.

An Unprecedented Visit

From the diary of John F. Fulton, professor, history of medicine, Yale University Medical School:

Wednesday, July 2nd [1941]—The hottest day of the year so far, with temperature at 94.5° during the afternoon in New Haven with very high humidity. . . . To my great surprise, about 10 p.m. in the evening, Howard Florey telephoned from New York, having arrived late this afternoon on the clipper with a medical assistant named Heatley, on a mission for the British Government and the [British] Medical Research Council. It [the trip] was evidently arranged rather hurriedly and he had managed to get high priority which had carried him through Lisbon with only two or three days delay. I persuaded him to come up the following afternoon to spend the long weekend with us [in New Haven] as he will be unable to get anything done over the holiday.

Thursday, July 3rd—. . . In the evening, [Dr. Stanhope] Bayne-Jones came in to discuss Howard's problem and his new disclosures concerning penicillin.

Friday, July 4th—Took Howard and Dr. Heatley to the Lab to look over N.R.C. [National Research Council] reports. At noon, Lucia [Fulton] had arranged her usual large Fourth of July cocktail party. More than a hundred people turned up, including [Drs.] Francis Blake, John Paul, and Jim Trask, all of whom were delighted to see Howard, as were the English mothers. [During World War II, Yale University faculty played host to and cared for a number of children from England, including Paquita and Charles Florey who lived for five years with Dr. and Mrs. John Fulton.]

Saturday, July 5th—Spent part of the morning with Professor Ross Harrison, who gave Florey and Heatley some very valuable leads with regard to their mission in this country. . . .

Sunday, July 6th—[Wrote] a letter to [Sir Edward] Mellenby concerning Howard Florey [which] gives the details, of his mission.[32]

Howard Florey and Norman Heatley arrived in the United States on July 2, 1941. The next day, before leaving by train for New Haven,[32,33] they paid their respects and reported the results of their study to Dr. Alan Gregg at the Rockefeller Foundation, which provided financial support for the trip. Four days later, they had been introduced to the proper people and their future course of action had been planned. All this happened in such a short time because of Florey's long-standing friendship with John Fulton who had come up to Magdalen at the same time as Florey and for three years had worked in the same laboratory with him in physiology under Sir Charles Sherrington.

Florey and Heatley had left for the United States within a few days after they had completed treatment of their first six patients. All were patients who had failed to respond to sulfonamide therapy and were severely ill with staphylococcal or streptococcal infections.[6,32,33] Although two ultimately died, all six initially showed impressive responses to the penicillin treatment. The May 25, 1940, experiment performed at Oxford with only eight mice had given Florey the courage to proceed with the treatment of these patients. That May 25 experiment was repeated several times before the drug was administered to the first patient. But it was that first in vivo chemotherapeutic trial that, despite the small number of animals used, convinced Florey, Chain, and their associates that somehow sufficient penicillin must be obtained to permit its evaluation in humans.

Norman Heatley deserves the credit for modifying the available laboratory procedures so that enough penicillin could be accumulated for this purpose. He supervised the entire production process and, most important, adapted to the study of penicillin an assay procedure introduced some years earlier for measuring the activity of lysozyme.[34–36] Without a way of quantitatively assaying the minute amounts of penicillin in culture media and at each stage of the extraction procedure, and without a defined unit of penicillin activity, little could have been accomplished.

Fleming and subsequent investigators previously had assayed the potency of penicillin-containing fluids by a serial dilution technique that measured the highest dilution of penicillin liquor that would prevent growth of the test organism. This procedure is still widely used for some purposes, but requires that solutions be sterile. Moreover, it provides only an estimate of the degree of activity.

The procedure Heatley recommended was based on diffusion of

penicillin through agar. Agar that had been inoculated with a standard penicillin-susceptible strain of *Staphylococcus aureus* was placed in a series of glass petri dishes and allowed to harden. Glass or porcelain cylinders were placed at uniform distances from one another on the surface of the hardened agar and were filled with the solutions to be tested. The plates were then incubated at 37°C until the staphylococci were fully grown. After incubation, clear circular zones of staphylococcal growth inhibition surrounded each cylinder containing active penicillin. The diameter of the zone was a function of the concentration of penicillin in the cylinder and could be quantified by reference to a standard curve. Greatly modified and refined, this procedure is still in use today.

The unit of penicillin introduced by Heatley was established as that amount of penicillin which when dissolved in 1 ml of water gives the same inhibition as an arbitrarily selected batch of penicillin (prepared at Oxford) which gives a 24 mm (average) zone of growth inhibition. Ultimately it became clear, when pure penicillin became available, that this unit was essentially equivalent to the biological activity of 0.6 mcg of crystalline sodium penicillin G (sodium benzyl penicillin). It was established in 1944 as the international unit of penicillin activity.[37,38]

Norman Heatley's development of a relatively quantitative assay procedure and his introduction of a defined unit of penicillin activity were probably the most important contributions made by anyone to the ultimately successful development of the drug. Heatley was a remarkably fine microbiologist and biochemist, and a most ingenious person. Minor difficulties were not allowed to become roadblocks. He knew that, if penicillin was to be produced in increased quantities, the first requirement was large numbers of vessels for growing the mold.[38,39] The usual conical flasks used for most microbiological work were wasteful of incubator and sterilizer space and were not available in sufficient numbers. Moreover, the Battle of Britain had only recently ended, and supplies were hard to obtain. He tried "bottles, trays, pie dishes, bedpans, plain or varnished tins, etc.," and eventually found that "the old-style bedpan with a side-arm and lid was an ideal culture vessel, providing a relatively large surface area over a shallow layer of fluid and with a side-arm for inoculation and withdrawal." Unfortunately, when an effort was made to obtain a sufficient number of these pans, it was found that they had been replaced by a more modern design with no lid.[38]

Glass vessels could not be obtained either within a reasonable period of time, and the mold (module) moreover would have been costly.

Through a friend of Florey, however, they obtained the assistance of James MacIntyre & Co., Ltd., of the Staffordshire pottery industry. The company agreed to supply the needed vessels, and Heatley went to its plant at Stoke-on-Trent to supervise their manufacture. By the time he arrived, they had already made some "unfired prototypes which the modeller was able to trim with the knife to a finally selected design." The first three, rectangular in shape, arrived in Oxford on November 18, 1940. A larger number was ready by December 23 and Heatley immediately went by van to obtain them. They were washed, filled, and sterilized on the following day and inoculated on Christmas Day.[38,39]

Soon after, the harvesting operation at Oxford was partially mechanized by a special trolley on which the rectangular vessels could be tilted so that their contents drained toward the spout, the medium could be sucked out through a replaceable sterilized tube, and replacement medium could be inserted.[38]

Vacuum distillation and later freeze-drying was used to concentrate the crude culture fluid. As shown previously by Clutterbuck, Lovell, and Raistrick in 1932,[40] the penicillin was completely removed from aqueous solution at pH 2 by shaking with ether, but it was only partially removed at pH 7.2. No active material was removed when the ether was evaporated alone, but, if evaporated over water, the residual water contained a considerable portion of the original activity. It was clear that penicillin passed more freely into ether at acid than at neutral pH. Clutterbuck and his associates had failed to deduce that penicillin must be back-extractable from ether into neutral buffer, and at Oxford this was recognized only "with difficulty." Lewis Holt had shown this in 1934 but unfortunately had not published his observations.[41,42]

The resulting solvent transfer method of extracting penicillin from cooled acidified crude culture fluid by shaking with ether and then taking it back from the separated ether phase into water at neutral pH was simple. It could be applied directly to unconcentrated culture fluids, and it freed penicillin from proteins, carbohydrates, and salts. Eventually, amyl acetate, being less volatile, less flammable, and less wasteful of solvent, replaced the ether and was used widely in the early development of penicillin. Amyl acetate had also been used by Lewis Holt in 1934 (see charts 3.3 and 4.1).[42]

Thus, largely through Heatley's ingenuity, methods were developed for the production of penicillin in quantities large enough for Howard Florey to initiate clinical trials in six patients at the Radcliffe Infirmary.

Heatley, incidentally, was not the only ingenious member of the Oxford team. A portion of the penicillin used to treat Florey's first six patients was derived from their urine after treatment. Howard Florey's wife, Ethel, a physician and clinical research investigator at the Radcliffe Infirmary in Oxford, later became a familiar sight as she rode about on her bicycle transporting the urine of treated patients to the laboratory for recovery of the penicillin.[43]

Only sixteen days after discontinuing treatment of the last of their six patients, Florey and Heatley arrived in the United States on their trip to spread the news. The trip, however, was not made in haste. As early as mid-April, before four of the six patients had even been admitted to the hospital or started on therapy, Florey had secretly begun to make plans. Somehow, larger supplies of penicillin had to be obtained.

The first patient suffering from an infection to be treated with penicillin parenterally was Constable Albert Alexander. He had been admitted to the Radcliffe Infirmary two months previously with suppurating abscesses resulting from a scratch on the face from a rosebush. "He had lost one eye, one humerus was involved, and neither surgical drainage nor treatment with sulfapyridine had controlled the mixed streptococcal and staphylococcal infection." Penicillin was administered by intravenous drip, beginning on February 12, 1941. Within four days, there was dramatic improvement. Unfortunately, however, supplies of penicillin ran out and after ten days the patient relapsed and died.[6,7,38]

The first [patient], treatment of whom was begun at the Radcliffe Infirmary on 12 February 1941, had a severe staphylococcal infection with abscess formation and osteomyelitis. At that time little was known about how long the treatment might have to be continued, and after 5 days, when considerable clinical improvement had taken place, the meagre stock of penicillin was exhausted and the patient eventually relapsed and died. The next, who was treated at the Wingfield-Morris Orthopaedic Hospital, Oxford, was a boy with osteomyelitis due to a haemolytic streptococcus, which had not responded to other treatment. Following the intravenous administration of penicillin for 5 days the infection appeared to be controlled, but about 3 weeks later...the infection flared up again.... The third patient was a man with a large carbuncle, which healed in a striking fashion after the administration of penicillin, without any further extension of the necrosis. The fourth case was one of cavernous sinus thrombosis due to a staphylococcus. Other therapeutic procedures had been tried and the patient, a 4-year-old boy, was considered to be moribund. The administration of penicillin was continued for 9 days in this case and led to a steady improvement. The boy was apparently restored almost to normal when he died on the 18th day of a ruptured mycotic aneurysm of the vertebral artery. At the autopsy it was found that the infection in the cavernous sinus, orbits and lungs had been almost entirely overcome and that the healing processes were well advanced. The fifth patient was another boy with staphylococcal septicaemia, accompanied by an early osteomyelitis, who had been treated without effect with sulphathiazole. He was the first patient to receive continuous intravenous penicillin and he made a good recovery with an excellent functional result after 10 days of continuous infusion and 4 days of intermittent injections. The sixth patient was a baby with a staphylococcal infection in the urinary tract, which cleared up rapidly after giving penicillin by mouth. The oral route was possibly successful here, although only small doses were given, because a baby has little acid in the stomach, and because an inhibitory concentration of the drug was only required in the urine and not in the blood. . . . From these cases it was clear that substantial doses of penicillin were not toxic to man—in fact one of the most striking features was the improved appetite and feeling of well-being of these gravely ill patients within 2 or 3 days of the beginning of treatment—and there were very good indications that the drug could control the most severe infections. The effect on staphylococcal infections was especially important, as the sulphonamides were of limited use against the staphylococcus."[6]

Florey and Heatley's purpose in coming to the States was not only to acquaint the medical profession with their findings but to interest the U.S. pharmaceutical industry in large-scale commercial production of penicillin, in time, they hoped, for its use in the treatment of war casualties. During 1940 and 1941, Hitler had focused Germany's war efforts on subjugation of the English. Operation Eagle, an all-out air assault on England, began on August 13, 1940, and the bombing of London, on August 23, the day before Florey and his associates' first report on penicillin appeared in the scientific literature. The bombing of London and its environs continued through the winter of 1940 to 1941, while Hitler planned for the "annihilation of Russia's vital energy." On June 22, 1941, Hitler launched the invasion of the USSR, and initially at least, he was successful in dominating the Russians.[44] Day-by-day mass extermination of Bolshevik leaders and Russian Jews followed, and what Hitler later called the "greatest battle in the history of the world" (referring particularly to the June 22 invasion) continued.[47–49]

The British pharmaceutical industry, therefore, could not help Florey at this time. It was hampered by inadequate personnel, limited materials for production, and the wartime demand for other products. Florey thought that the Americans might be in a better position to help. So he and Heatley arrived in the United States with high hopes of arranging with U.S. scientists a collaborative project directed toward large-scale manufacture of the drug. Their hopes for this venture were soon dashed, however, by the Japanese attack on Pearl Harbor on December 7, 1941, and the United States' subsequent declaration of war against Japan on December 8 and against Germany and Italy on December 11 (after Germany and Italy had declared war against the United States). Close collaboration between scientists in England and those in the United States was virtually impossible from then on. Florey nevertheless had accomplished his mission. By December 1941, interest had been aroused, and many were already feverishly working on all aspects of the penicillin problem.

5

Birth and Development of an International Collaborative Program

Howard Florey and Norman Heatley met Ross G. Harrison in New Haven a few days after their arrival in the United States. Harrison, who was chairman of the National Research Council (NRC),[1] was in a position to be of major assistance to the two visitors from England. He suggested that they see Dr. Charles Thom who had long been involved in studying the penicillia. As we saw earlier, it was Thom who, at the request of Dr. P. W. Clutterbuck, had identified Alexander Fleming's "contaminant" as a strain of *Penicillium notatum* Westling. Florey and Heatley immediately went to Washington to meet Thom who was then affiliated with the Bureau of Plant Industry of the U.S. Department of Agriculture.[2,3]

Again an element of chance entered the penicillin story. Percy A. Wells, director of the Eastern Regional Research Laboratory of the Agricultural Research Service (U.S. Department of Agriculture), was serving as acting assistant chief of the Bureau of Agricultural and Industrial Chemistry.[4] In July 1941, he was temporarily in charge of the four regional research laboratories that had been authorized by Congress in February 1938 to develop new uses for farm products. Thom took his two guests to see Wells and told him of their mission—to obtain increased supplies of penicillin. Because mold fermentation was his special interest, Wells promptly recognized the solution to the problem. Improving yields of products had been one of the principal objectives in every fermentation process he had studied, and usually he and his associates were successful.

The Northern Regional Research Laboratory (NRRL) in Peoria had opened late in 1940, just in time, as it turned out, to undertake Florey's penicillin project. Wells knew that a special pilot-type shallow-pan alumi-

PLATE 5.1.   *Penicillin research team at the Northern Regional Research Laboratory, June 1944: (foreground left) William Schmidt, microbiologist in charge of penicillin assays; Max Reeves, laboratory technician; Morris Friedkin, assistant to Dr. Stodola, later became a well-known enzyme chemist; Jacques Wachtel, chemist; (foreground right) Z. Louise Smith, chemical technician, associated with Dr. Ward; Dr. L. J. Wickerham, specialist on yeasts; George Nelson, chemical technician who worked initially with Dr. Ward and is still located at the NRRL; (background, left to right) the late Dorothy Alexander Fennel, microbiologist; the late H. T. Herrick, then director of the NRRL. Herrick's team provided the nucleus for Coghill's research team at the NRRL during World War II; F. H. Stodola, analytical chemist, who prepared (from pooled contributions of penicillin G) the penicillin International Standard that is still used today; Kenneth Raper, chief mycologist; Dr. Robert Coghill, director of the fermentation division of the NRRL; Dr. George Ward, prime associate of Coghill; Dr. Andrew Moyer, who jointly shared with Coghill the patent covering the addition of phenylacetic acid to culture media to secure high yields of penicillin G. Moyer also secured a patent on the use of corn steep liquor in culture media to increase penicillin yields and licensed the foreign rights to the Commercial Solvents Corp.; Dr. Reid T. Milner, chief of the analytical division of the NRRL; Dr. N. C. Schieltz, physiological chemist; Dr. C. H. Van Etten, microanalytical chemist, who analyzed the International Standard.*

num fermentor had been constructed not long before and might be available—"probably still uncrated"—at the Peoria laboratory. He felt that the fermentor might well be useful in the penicillin project. Accordingly, he telegraphed Dr. Orville E. May, director of the NRRL, explaining Florey's request and asking if in fact the equipment was available.

C216 41 DL GOVT=PA WASHINGTON DC 9 309P 1941 JUL 9 PM 3 29 DR OE MAY=

NORTHERN REGIONAL RESEARCH LABORATORY 825 NORTH UNIVERSITY PEORIA ILL=

THOM HAS INTRODUCED HEATLEY AND FLOREY OF OXFORD, ENGLAND, HERE TO INVESTIGATE PILOT SCALE PRODUCTION OF BACTERIOSTATIC MATERIAL FROM FLEMING'S PENICILLIUM IN CONNECTION WITH MEDICAL DEFENSE PLANS. CAN YOU ARRANGE IMMEDIATELY FOR SHALLOW PAN SETUP TO ESTABLISH LABORATORY RESULTS IN METAL=

P A WELLS.[4,5]

Within a half hour, Wells had done what he could for Florey and Heatley.

A telegram came from Dr. May on the following morning inviting Florey and Heatley to visit him and Dr. Robert Coghill, Chief of the Fermentation Division of the NRRL, in Peoria, Illinois.

JULY 10, 1941 8:50 AM (TELEGRAM TO WELLS.)

PAN SETUP AND ORGANISMS AVAILABLE HEATLEY AND FLOREY EXPERIMENTATION DETAILS OF PROPOSED WORK OF COURSE UNKNOWN AND SUGGEST THEY VISIT PEORIA FOR DISCUSSION. LABORATORY IN POSITION TO COOPERATE IMMEDIATELY.

O. E. MAY
NORTHERN REGIONAL RESEARCH LABORATORY[4,5]

Wells replied on the same day:

OEM:mgm
STRAIGHT TELEGRAM
COLLECT

JULY 10, 1941 (MEMO TO MAY FROM WELLS.)

THANK YOU VERY MUCH FOR YOUR TELEGRAM THIS MORNING, INDICATING THAT THE NORTHERN LABORATORY IS IN POSITION TO COOPERATE IMMEDIATELY WITH PROFESSORS HEATLEY AND FLOREY ON THE PILOT-SCALE PRODUCTION OF BACTERIOSTATIC MATERIAL FROM FLEMING'S PENICILLIUM....

I KNOW IT WILL OCCUR TO YOU AND TO THOSE IN THE FERMENTATION DIVISION TO TRY OUT THE PRODUCTION OF THIS BACTERIOSTATIC AGENT IN SUBMERGED CULTURES. (MAY AND OTHERS PATENTED A SUBMERGED FERMENTATION PROCESS FOR USDA IN 1935 AT ARLINGTON FARMS, VIRGINIA, NOW SITE OF THE PENTAGON.

WELLS WAS ON MAY'S STAFF.) THIS WOULD MOST CERTAINLY FACILITATE ITS
COMMERCIAL PREPARATION.

P A WELLS[4,5]

Florey did not know at the time that the NRRL was probably the only laboratory in the United States where it was routine practice, whenever a fermentation problem was assigned to them, to try corn steep liquor in the culture medium. Nor did he know that the use of corn steep liquor would be so critically important to the outcome of the penicillin project.

The use of corn steep liquor in fermentation media dated from 1925, even before the existence of the NRRL. Robert Coghill has commented, "The discovery of corn steep liquor's key place in penicillin fermentation media was foreordained and inevitable," once the problem of increasing yields of penicillin was assigned to the NRRL's Fermentation Division.[4-6]

Florey and Heatley arrived in Peoria on July 14. The first meeting was held that day, and work directed toward the production of penicillin began on the following day. Under the leadership of Robert Coghill, who had moved from Yale University to the NRRL in 1939 to organize its Fermentation Division, the NRRL staff continued to work on penicillin for four years. It is safe to say that most if not all of the major biological contributions to penicillin's large-scale production emanated from that group.[6]

Howard Florey stayed in the United States until the middle of September, visiting university and pharmaceutical company laboratories to try to arouse interest in the production of penicillin. Assured that steps would be taken in this direction, he finally returned to Oxford to get on with his own studies of the drug, and await impatiently first supplies of the substance from the United States.

Norman Heatley stayed in Peoria until November 30. By this time he had passed on to the others full information concerning his penicillin assay procedure. Moreover, Dr. A. J. Moyer by this date had already introduced corn steep liquor into the penicillin fermentation liquor and increased yields more than tenfold (from 1 to 2 units per milliliter to approximately 20 to 24 units per milliliter). Heatley then affiliated with Merck & Co., Inc., in Rahway, New Jersey, where he worked for another six months, sharing the knowledge acquired at Oxford and trying to help with early phases of the penicillin project. Unaccustomed to a large industrial laboratory, however, he was often frustrated at Merck by all that went on around him and all that he was not a part of. He returned to England to

PLATE 5.2.   Robert D. Coghill, Ph.D., chief, Fermentation Division, Northern
Regional Research Laboratory, U.S. Department of Agriculture.

work thereafter on cephalosporin and other antibiotics as well as penicillin.[6,7]

Florey had asked the United States only for enough penicillin to treat "another score or more of patients." He received far more, but only after what seemed to him endless delays.

Penicillin Research and Development Implemented by the United States Government

Research on penicillin, its production, and its clinical evaluation in the United States, during 1941 to 1945, was under the U.S. Office of Scientific Research and Development (OSRD) and the U.S. War Production Board (WPB).[8-11] Neither of these administrative organizations came about as the result of Howard Florey's visit to the United States. Rather, they had been planned and set up previously—soon after it was clear that Great Britain and probably the United States would be drawn into a war, the magnitude of which had not been experienced before.

Under the National Defense Research Committee, created in late 1940, the Office of Scientific Research and Development (the OSRD) was created in June 1941

to assure adequate provision for research on scientific and medical problems relating to the national defense. The office was terminated on December 31, 1947. . . . During the war period the Office served as a center for the mobilization of scientific personnel and resources of the Nation and it cooperated in planning, aiding, and supplementing, where necessary, the experimental and other research activities carried on by the armed services and other federal agencies. To this end, it was authorized to enter into contracts and agreements with individuals, educational and scientific institutions (including the National Academy of Sciences and the National Research Council), industrial organizations, and other agencies for studies, experimental investigations, and reports.[8]

Dr. Vannevar Bush, who had been chairman of the National Defense Research Committee since 1940, served as director of the OSRD. Dr. A. N. Richards of the University of Pennsylvania served under Bush as chairman of its subdivision, the Committee on Medical Research (CMR). Civilian members of the CMR were Dr. Lewis H. Weed of Johns Hopkins University and the National Research Council, Dr. Alphonse R. Dochez of Columbia University, and Dr. A. Baird Hastings of Harvard University. Brig. Gen. James Stevens Simmons (M.C.), U.S.A.; Rear Admiral Harold W. Smith (M.C.), U.S. Navy; and Dr. L. R. Thompson (later replaced by Dr. R. E. Dyer) of the U.S. Public Health Service were also members of

PLATE 5.3. *Prof. Alfred Newton Richards, M.D., vice-president, University of Pennsylvania, and chairman of the U.S. Committee on Medical Research under the Office of Scientific Research and Development during World War II. (Courtesy of the Department of Pharmacology, School of Medicine, University of Pennsylvania.)*

the committee. Dr. Irvin Stewart of the Carnegie Corporation of America served as its executive secretary.[9]

The first contribution of the CMR to research on penicillin was to make funds available to the Department of Agriculture's Northern Regional Research Laboratory in Peoria in the summer of 1941. In February 1942, it formalized the arrangement through a contract with the Bradley Technologic Institute (also in Peoria), with Dr. Robert Coghill as responsible investigator. Funds were then made available on a continuing basis for penicillin studies at the NRRL. Later the CMR assumed responsibility through the National Research Council Committee for a clinical evaluation of the drug under the supervision of Dr. Chester Keefer, and still later, under the scientific leadership of Dr. Hans T. Clarke of Columbia University, for contracts with various firms and colleges to study the structure of penicillin. The ultimate goal of the National Research Council was the chemical synthesis of the substance.

The OSRD exchanged information directly with British ministries through one of its principal subdivisions, the Liaison Office in London, or the London Mission. Its authority to enter into contracts gave the OSRD its effectiveness, flexibility, and power. In the case of the penicillin project, its strength derived from its ability to demand results at the production level, to control distribution, and to effect clinical trials that produced irrefutable information. Also important was the fact that its subcommittees were identical in name, purpose, and composition with those established simultaneously by the National Research Council, the active arm of the National Academy of Science.[10,11]

The NRRL and Penicillin

Research on penicillin began at the NRRL on July 15, 1941. No time was devoted to the long discussions, literature reviews, and careful design of protocols that today would normally precede initiation of a new research project. H. T. Herrick, who was assistant chief of the Bureau of Agricultural and Industrial Chemistry of the U.S. Department of Agriculture, was responsible for the operations of four regional research laboratories authorized by Congress in February 1938 to develop new uses for farm products. He and his associates at the NRRL, Dr. Orville E. May, director, Dr. Robert Coghill, head of its Fermentation Division, Dr. G. E. Ward, Coghill's deputy, and Dr. Kenneth B. Raper and Dr. Andrew J. Moyer, had been through this before. They had not produced penicillin previously, but they knew ways of encouraging penicillia and other molds to produce high yields of their metabolic products.[2-4]

The agreement reached between the NRRL and Florey and Heatley was that (1) the NRRL would try to develop a culture medium that would favor production of larger amounts of penicillin; (2) it would try to find or develop strains of penicillia capable of producing larger yields of penicillin than had been obtained at Oxford; and (3) it would simultaneously investigate the possibility of producing penicillin by submerged growth of the organism.[4-6]

Florey and Heatley had brought to Peoria freeze-dried cultures of their penicillin-producing mold, a few of the glass and porcelain cylinders used at Oxford for assay of penicillin-containing solutions, and a small amount of brown powder containing 42 Florey (later called Oxford) units of penicillin activity per milligram. The powder was to be used in biological assays as a standard for comparison of the activity of penicillin made in the United States with that made in England. Florey gave assay cylinders and small samples of the penicillin standard powder to the NRRL, to industrial laboratories, and to some other laboratories including ours at Columbia University and Wallace Herrell's at the Mayo Clinic in Rochester, Minnesota. Heatley gave each laboratory detailed information on the assay procedure he had developed. Thus a direct tie between penicillin research in England and that in the United States was ensured.

The penicillin-producing microorganism is a saprophyte belonging to the genus *Penicillium*. Originally described by Alexander Fleming as a strain of *Penicillium rubrum* Biourge and later identified by Thom as a strain of *Penicillium notatum* Westling, it was assigned the number 144.5112.1 in Thom's collection of strains. The culture brought to the United States by Florey and Heatley in 1941, although presumably the same as that received by Thom for identification in 1930, was designated 144.5767 when received in Thom's collection.[2] It was thought at first that penicillin production was a property of the Fleming strain only. It was soon realized, however, that the substance could be produced, in greater or lesser quantities, by most if not all strains of the chrysogenum-notatum group of penicillia.

Penicillia are aerobes and grow as a scum or a feltlike layer of mycelium on the surface of a fluid nutrient solution, or they may form a velvety green mat on solid or semisolid nutrient materials.[2,5,6,12-15] Growth starts from a spore that germinates and produces septate hyphae, which in turn grow and branch, forming a thick felted colony. The hyphae

95

may penetrate one or two millimeters below the surface of the nutrient medium, but usually they go no deeper because they need air to survive. The hyphae subdivide and form bodies, each of which resembles a brush or pencil (a penicillus). The spores bud off from the terminal branches of these reproductive hyphae and may be transmitted on currents of air. The possible transmission of these spores in air was of considerable importance to investigators concerned with other fermentation processes or with the commercial production of other substances requiring absolute sterility.

Strains of *Penicillium notatum* are related to but not identical with those penicillia often seen as contaminants on bread or those used in the manufacture of some cheeses. The thick felt produced by growth of the septate hyphae, when covered with spores, has a deep gray-green color. Yellow droplets containing pigment and high concentrations of penicillin characteristically may be seen on the surface of the green felt.

The aerobic nature of the organism led Fleming and subsequent investigators prior to 1941 to grow it in "stationary" or "still" culture—in shallow layers usually not more than one inch in depth—when attempting to produce penicillin. The mold's inability to penetrate more than one to two millimeters into the medium meant that enormously large surface areas of medium were necessary for production of any significant amount of penicillin. The number of culture flasks (bottles or trays) required for the process was tremendous. This was a major problem and led some to believe that penicillin would never be produced in amounts sufficient for general clinical use unless it could be grown submerged in vats or tanks. John L. Smith, vice president of Chas. Pfizer & Co., manufacturer of fine chemicals many of which were produced by fermentation processes, insisted on this from the beginning.[16] Gluconic, lactic, and fumaric acids already were being produced by Pfizer in submerged culture, and from 1941 on this company among others directed its attention primarily to the development of methods for penicillin's production in deep tanks.

Yet most of the penicillin that became available for clinical use prior to May 1943, from Pfizer as well as other companies, was produced in shallow layers—in flasks, bottles, or pans.

At Oxford a boiled extract of brewer's yeast was an important component of the culture medium used for penicillin production. At the NRRL, this was not available, and Andrew Moyer suggested that corn steep liquor, a by-product of the wet corn milling industry, be tried as a . substitute for the yeast.[6] At the time, even Moyer may not have known the importance of this suggestion.

Corn steep liquor was cheap, was readily available to the NRRL, and

PLATE 5.4.   Penicillium chrysogenum *showing pigmented (yellow) droplets of penicillin exuding from surface of the mold growth. (Courtesy of Pfizer, Inc.)*

*PLATE 5.5.   John L. Smith, vice-president, later president, Chas. Pfizer & Co.,
about 1946.*

was already used there as a source of growth factors in certain other fermentations. Quickly it became apparent that it not only accelerated growth of the penicillin-producing strain of *Penicillium notatum*—as did the yeast extract used in England—but also increased the yield of penicillin. Moreover, further increases in yield occurred when the trace metal and carbohydrate compositions of the medium were adjusted and lactose was substituted for glucose.

Corn steep liquor, a by product of the manufacture of corn starch, was available in great quantity in the vicinity of Peoria, located in the corn belt of the United States. It was known that maize contained a protein, zein, which was rich in phenylalanine; and it was assumed that corn steep liquor also might contain phenylalanine and some metabolic derivatives of the substances produced from it by bacterial action. Extensive study of the composition of corn steep liquor was carried out by Moyer and Coghill at the Northern Regional Research Laboratory in Peoria, by scientists at the Corn Products Refining Company, and by others. Ultimately it became evident that the liquor contained phenylacetic acid derivatives which specifically stimulated the production of benzyl penicillin. This was the first use of precursors to increase penicillin production. Later various other substances were used to direct the fermentation toward the formation of specifically desired forms of penicillin.[6] Unfortunately, the British initially were unable to profit from the use of corn steep liquor, for corn was not grown in England.

Development of the corn steep liquor–lactose medium used at the NRRL is credited as one of the most important developments leading to the successful large-scale production of penicillin. It stands second only to Heatley's development of a quantitative procedure for assay of penicillin solutions. Of almost equal importance was the observation made at Peoria in 1941 that, when their corn steep liquor–lactose medium was inoculated with a penicillin-producing strain of *Penicillium notatum* and subjected to continuous vigorous agitation in revolving drums (as used for production of gluconic acid and in certain other fermentation processes), the organisms would grow, submerged in the medium, in the form of small rounded pellets and would secrete penicillin into the medium.

The isolation and/or development of better penicillin-producing strains of penicillia was the fourth major step that made large-scale production of the antibiotic possible. These strains, however, were not obtained quickly.

All early studies on penicillin in both the United States and England were performed with Fleming's original strain, designated in the United

States as NRRL 1249. Yields approximated 2 to 4 units of penicillin per milliliter of fermentation liquor. By single-spore isolations and repeated testing of the resulting clones, a substrain (designated NRRL 1249.B21) was developed at Peoria that allowed yields up to 180 to 200 units per milliliter of culture medium. NRRL .B21 was widely distributed; for some time it was used extensively for surface fermentation of penicillin. Later, a strain of *Penicillium notatum*, genetically unrelated to Fleming's strain, was found in the NRRL culture collection. This strain—NRRL 832, originally from Biourge's laboratory in Belgium—when grown submerged, produced lower yields than strain NRRL .B21 produced in surface culture. But the large volumes of liquor in vats or tanks, even though containing only 40 to 50 units per milliliter, allowed greater overall production than smaller volumes of surface-grown culture fluids containing up to 200 units per milliliter. The strain NRRL 832, moreover, produced primarily penicillin G, a relatively stable form of the antibiotic, whereas strain NRRL .B21 primarily produced a less stable form, penicillin F.

Dr. Kenneth Raper, working with Dr. Robert Coghill at the Northern Regional Research Laboratory in Peoria, attempted to isolate from soil, organisms that would produce large amounts of penicillin. Coghill had made contact with the Army Transport Commmand, and samples of soil "in bottles, paper bags, paper cartons" came to them daily from all parts of the world. Good penicillin-producing strains were isolated from soil samples from Cape Town, Chungking, and Bombay, among other places. Ironically, however, the best strain came from Peoria.

This important strain was isolated from a moldy cantaloupe from a market in Peoria. A strain of *Penicillium chrysogenum*, designated NRRL 1951, it yielded a mutant (NRRL 1951.B25) that produced high yields of penicillin, particularly when cultivated under submerged conditions. An X-ray mutant (strain X-1612) of the NRRL 1951.B25 strain produced more than 500 units of penicillin per milliliter of medium. Spores of the X-1612 strain, when exposed to ultraviolet irradiation, yielded a strain (designated Q-176) that in turn produced more than twice that produced by X-1612 but largely in a less desirable form, penicillin K. By the addition of phenylacetic acid to the medium, known by then to be a penicillin precursor, the process could be redirected, however, and the high potencies of penicillin G retained.[11-13]

The strain X-1612 was isolated at the Carnegie Institution by Milislav Demerec, who showed that when an antibiotic-yielding organism is subjected to controlled dosage with ultra-short-wave radiation at an intensity insufficient to kill the organism but sufficient to produce chromosome rupture in the cell, mutants are developed that may be capable of producing extremely large amounts of the antibiotic. Q-176 resulted from ultraviolet irradiation studies at the University of Wisconsin.[13-15,17]

Three to four years were required for these high-yielding strains to evolve, and the development of penicillin as a useful product was accomplished without knowing how great the potential of strain selection was.

Before Heatley left Peoria, he had trained the staff at the NRRL to carry out penicillin assays by his method. Further, the yield of penicillin in fermentation liquors had been raised from 1 to 2 units per milliliter (as commonly obtained at Oxford) to approximately 20 to 24 units per milliliter. On December 2, Moyer and Heatley reported to Robert Coghill that the "study has resulted in a greatly improved method for penicillin production whereby yields are increased twelve-fold over those previously reported. The essential features of the new method are the additions of corn steeping liquor and of neutralizing agents to the medium." These yields were obtained by surface growth of the organism within five days' incubation. Many variations in media composition, temperature, and other environmental conditions had been tested, as well as growth of the organism in rotary drums and vats.[6,8]

Abraham and Chain had shown in 1940 that an extract of *Escherichia coli* cells of a strain insensitive to the antibacterial action of penicillin would completely inactivate the substance. Since the inactivator had the properties of an enzyme, they named it *penicillinase*.[18] Subsequently it was established by others that penicillinase is produced by a wide variety of organisms resistant to penicillin including penicillin-resistant organisms residing in vivo; and in recent years, penicillinase has become an important factor in clinical medicine, contributing to an increase in the number of patients not responding to treatment with penicillin. In the early 1940s, however—at the NRRL and elsewhere—penicillinase was of interest primarily because of the frequency with which organisms producing it were present as contaminants in penicillin-producing cultures. Penicillin production was seriously hampered by the ubiquitous nature of these contaminants.

On December 17, 1941, Coghill reported on the results of Moyer and Heatley's work at a meeting at the University Club in New York, called by Dr. A. N. Richards, chairman of the Committee on Medical Research of

PLATE 5.6.   Dr. Andrew Moyer and Dr. Norman Heatley at the Northern Regional Research Laboratory, 1941. (Courtesy of Dr. Robert Coghill.)

PLATE 5.7.   *George Merck, president, with Dr. Randolph Major, director of research, Merck & Co., April 1951.*

the Office of Scientific Research and Development, and attended by Dr. Baird Hastings and Mansfield Clark of the National Research Council and representatives from Merck & Co., E. R. Squibb & Sons, Chas. Pfizer & Co., and the Lederle Laboratories. The report convinced those present at the meeting, particularly George Merck, that mass production of penicillin might be feasible.[6,8]

Steps Toward Establishing Penicillin as a Useful Drug

By December 1941, the studies at Oxford, at the NRRL, at Columbia University, and at a few other laboratories had made clear penicillin's

103

remarkable antimicrobial activity under experimental conditions. But many questions remained unanswered. Would penicillin be as active in the treatment of infections in humans as it appeared in vitro and in the treatment of experimental infections in animals? Would it produce in large numbers of patients the same therapeutic responses that Florey had observed in the six patients he had treated before coming to the United States? Was its spectrum of activity broad enough, its degree of activity great enough, to justify the vast expenditure of monies necessary to effect the transition from small-scale research to commercial production? Once available in quantity, would there be a sufficient market for the substance to justify its continued large-scale production?

Contacts made by Florey and Heatley while in the United States and Dr. A. N. Richards's interest in helping Florey led to the United States' undertaking to answer these questions. In his capacity as chairman of the Committee on Medical Research, Richards urged a few carefully selected pharmaceutical companies to embark on extensive research programs to establish if penicillin really could be mass produced. In the meanwhile, they were to produce enough penicillin so that he might establish its value in the treatment of infections that were likely to occur in the armed services and the civilian population. Dr. Richards wrote later:

The progress made by the Oxford team and the results which Florey presented convinced me that a collaborative effort such as Florey envisaged should be undertaken. A decision to that effect was approved by Dr. [Vannevar] Bush and formally acted on by the C.M.R. at its sixth meeting on October 2, 1941. A resolution was passed that "the Chairman be authorized to suggest to interested persons the desirability of a concerted programme of research on penicillin involving the pooling of information and results; and if the responses are favourable, to proceed to arrange for a conference on the subject.[19]

The first conference was held on October 8, with Dr. Vannevar Bush presiding.[19] Others attending were Dr. Richards, Dr. Lewis H. Weed (vice chairman of the CMR), Dr. William Mansfield Clark (chairman of the Division of Chemistry of the National Research Council), Dr. Charles Thom (chief mycologist of the Department of Agriculture), and the research directors from four commercial firms "thought to be capable of producing penicillin"—R. T. Major of Merck and Co., George A. Harrop of E. R. Squibb and Sons, J. H. Kane of Chas. Pfizer and Co., and Y. SubbaRow of Lederle Laboratories. Each of the four companies had its own degree of expertise in the necessary production processes. Merck

PLATE 5.8.   Dr. George Harrop, first director of the Squibb Institute for Medical
Research. He demonstrated that fundamental research can proceed within an
industrial organization.

perhaps had done the most research up to that time on penicillin
production. Pfizer had the most experience in large-scale fermentation
processes and for some years had commercially produced citric, lactic,
tartaric, and gluconic acids by fermentation. Pfizer was experienced too in
commercial production by both surface and submerged growth of microorganisms.

While the U.S. government was encouraging industry to study penicillin on a large scale and to produce—even before the research was
completed—enough to allow its clinical evaluation, small laboratories

105

throughout the United States and England and an increasing number of commercial firms became interested. They knew that the chemotherapy of bacterial infections had become a reality when the sulfonamide drugs were introduced in 1935. They also knew that by 1939, even among those bacterial species most susceptible to the sulfonamides, partially or fully resistant microbial strains had been observed. Effective as the sulfonamides were, such observations represented a signal that their efficacy might be short-lived.

From the beginning, penicillin was obviously the most active and most highly selective antimicrobial ever described. It was active against sulfonamide-susceptible and sulfonamide-resistant strains and seemed to produce no toxicity in the mammalian host. It promised to do all that had been expected of the sulfonamides.

In the months that followed, many rushed to become a part of the penicillin development program, to study methods for its fermentation and extraction, and to learn about its chemical, physical, and biological properties. Penicillin was usually produced as a sodium or ammonium salt (later as the calcium salt also), an orange-brown hygroscopic powder with a slight musty odor. The impure product was soluble in water and in alcohol, although somewhat unstable in the latter solvent. In the dry form it was stable to light but adversely affected by heat. At temperatures below 10°C (as in a refrigerator), it retained its full activity for several months and was obviously more stable than Fleming had suspected ten to twelve years earlier.

The scope of penicillin's activity was first defined by Fleming in 1929,[20] later by Clutterbuck et al.,[21] Roger Reid,[22] Chain and his associates,[23,24] and by Dawson, Meyer and myself.[25] It was easy to demonstrate its powerful action against gram-positive organisms, both aerobic and anaerobic, although it was the substance's growth-inhibiting properties—not its ability to lyse microbial cells—that we demonstrated. It was clear that it had only weak, if any, activity against most gram-negative microorganisms and offered little promise for use in the treatment of typhoid fever, dysentery, and infections caused by gram-negative bacilli. Everyone agreed that penicillin acted either as a bactericidal or bacteriostatic agent, depending on experimental or environmental conditions.

In our laboratory in 1940 and 1941, we cultured pneumococci, streptococci, and staphylococci with penicillin under various conditions in order to evaluate the factors that might influence the rate of killing by penicillin. By enumerating the surviving organisms at various times during the organisms' contact with the antimicrobial, we were easily able to estimate the rate of killing. The number of survivors decreased by geometric units as time increased by arithmetic units, and the log of the number of survivors plotted against time followed a straight line until at least 99 percent of the organisms were killed. Some viable cells always persisted, however, a fact later found to be of clinical significance. Similarly, the rate of killing by penicillin decreased as the number of organisms in the original inoculum increased and, within limits, increased as the concentration of antimicrobial increased. Throughout, penicillin exerted demonstrable antimicrobial activity only against organisms that were in the active phase of growth. The bactericidal action of penicillin contrasted sharply with the bacteriostatic effect of the sulfonamides, which were able to slow rates of growth but failed to kill susceptible organisms.[26] All this later proved important in studies on penicillin's mechanism of action, important too in attempts to establish what might have led to Fleming's discovery in 1928.

No detectable penicillin was absorbed onto microbial cells or was destroyed during the process of bacterial growth inhibition. Moreover, the action of penicillin was not diminished by the presence of serum, whole blood, or tissue autolysates. It even seemed to us that at times penicillin's activity was enhanced, at least in vitro, by the presence of body fluids. Since all our studies were performed with pathogenic organisms that grew more rapidly in the presence of body fluids, the apparently greater action of penicillin probably was to be expected.

First evidence that resistance to the action of penicillin might become a problem was provided by Abraham and his associates at Oxford in 1941.[24] By growing a highly susceptible strain of *Staphylococcus aureus* in vitro in progressively increasing concentrations of penicillin, resistance of the strain was increased 1,000-fold. A year later, Clara McKee and Dr. Geoffrey Rake at the Squibb Institute for Medical Research doubled the resistance of two strains of pneumococci by repeated transfers in broth containing the highest concentration of penicillin that would permit growth of the organism, but reversion to full susceptibility occurred on inoculation of these strains into animals.[27] In 1942 to 1943, McKee and her associates observed that 60, 29, and 28 serial passages through subinhibitory concentrations of penicillin increased the resistance of three strains of staphylococci 6,000-, 1,000-, and 4,000-fold, respectively. They found it easier to induce resistance in staphylococci than in pneumococci or hemolytic streptococci. Virulence of their strains for experimental animals seemed to decrease, moreover, as penicillin resistance increased.[27,28]

Of special importance were the observations of Dr. Leon Schmidt and C. L. Sesler, which showed that penicillin resistance could be induced in vivo while infected animals were receiving treatment with the drug.[29] They exposed two strains of pneumococci to penicillin by serial passage through seven groups of mice that were receiving injections of penicillin. After the sixth transfer, penicillin resistance of both strains was apparent when the drug failed to afford protection to the infected animals.

As Howard Florey and his associates at Oxford had done in 1940, most in vitro studies of penicillin's action performed subsequently by others were paralleled by protection tests in experimental animals.[5,30-35] These indicated that penicillin was highly active in vivo as well as in vitro. Consistent with the in vitro degree of susceptibility of the organisms tested and the number of Oxford units of penicillin administered, penicillin routinely protected infected animals from death when the drug was administered intravenously, intraperitoneally, or subcutaneously. Treatment was effective even when started long after the animals were infected. Certain aliphatic esters of penicillin, moreover, although inactive in vitro, similarly afforded considerable protection to animals. The discrepancy between the effects in vitro and in vivo was readily explained by the slow hydrolysis of the inactive ester within the animal and the resultant slow release of active penicillin. Unlike penicillin itself, moreover, its esters, by virtue of their greater stability at the pH of gastric juice, were active when administered by mouth.

The chemotherapeutic effects of penicillin were soon demonstrated against many other species of disease-producing microorganisms, including those strict anaerobes that cause gas gangrene and the spirochete of relapsing fever.[31-35] The activity of penicillin against gas gangrene and against the spirochete of syphilis proved of major importance, particularly during the World War II period.

Perhaps because of the apparent innocuous nature of the drug, few early attempts were made to study its pharmacological action. The current sophistication in the evaluation of new drugs (at least new antimicrobial drugs) was nonexistent at that time. Acute toxicity studies in mice revealed only that the lethal dose of intravenously administered penicillin was related directly to the purity of the penicillin preparation, the penicillin salt used, and the amount of sodium, potassium, or calcium ion present.[36]

All data tended to indicate that penicillin possessed only mild cytotoxic properties. Florey and his associates at Oxford had reported that leucocytes were immediately killed by high but not by low concentrations of their earliest penicillin preparations.[5] Similarly, Medawar concluded that there was little or no cytotoxicity due to penicillin, although the actual amount of drug used in his studies with fibroblasts of the chick embyro heart was not known. Later, Dr. Wallace Herrell and Dorothy Heilman at the Mayo Clinic in Rochester utilized inhibition of the growth and migration of macrophages and such tissue elements and of cells of the mesenteric lymph node of rabbits as their criterion of cytotoxicity. They concluded that penicillin was approximately one-tenth as toxic by weight as gramicidin, an antibiotic discovered by Dr. René Dubos in 1939 and already being used to some extent clinically, when tested in the same preparation containing explants of lymph nodes.[37,38]

Wallace Herrell's tissue culture studies with penicillin following Henry Dawson's presentation to the Young Turks on May 5, 1941, laid the foundation for the later clinical studies of penicillin at the Mayo Clinic. Herrell and Heilman reported the results of their studies at the December 1941 annual meeting of the Society of American Bacteriologists. The scientific community, the press, and the public showed great interest in what was probably the first symposium on antibiotics held in the United States. Since May 1941, each new development relating to penicillin had been widely publicized. This continued through 1942 and indeed until the end of World War II. The publicity in magazines and newspapers put strong pressure on the government and the pharmaceutical industry, and it also created a public demand for penicillin that had to be satisfied.

Why did the U.S. pharmaceutical industry hesitate to gamble on this new product? Fear—that the penicillin-producing mold would contaminate other products on which they depended for sales and revenues. Fear—that the ease of contamination of penicillin-producing cultures, the instability of penicillin in the presence of an enzyme produced by some of these contaminants, and the still too low yields of penicillin then obtained in culture fluids would lead to excessive production costs. Fear—that penicillin would be synthesized so rapidly that equipment installed in 1942, even in 1943, would soon be obsolete. Fear—that the I. G. Farbenindustrie and Farbwerke Hoechst in Germany, with their remarkable ability to synthesize such compounds as "606," prontosil and related compounds, would take control of the market. Fear—that the substance would prove to be more toxic or less effective when used in large numbers of persons than was apparent in the few patients treated prior to December 1941.

Each company knew that the cost of developing equipment and techniques for commercial production of penicillin would be so great that, once undertaken, the venture would have to be successful from the beginning. Each knew though that penicillin must be produced in significant quantity if its clinical value was to be established. They knew too that they must determine whether or not penicillin would really alter the course of infectious diseases in humans.

It is difficult now to document what transpired at that time. Research and production went hand in hand. Batches of penicillin made for research purposes often provided enough to help meet production quotas, and production units often supplied information that normally would be acquired through small experiments carried out as part of a research program. New procedures often were tested for the first time during the course of production runs. This was a gamble, for costs were high when experiments failed.

Producing sufficient penicillin to establish its merit was a slow process. The first small amounts of commercially produced penicillin distributed under the OSRD became available in the United States in March 1942. By the end of the year, enough had been produced to treat only one hundred patients. The many difficulties encountered in producing the quantities needed for therapeutic trials led early in 1942 to an agreement between penicillin producers and the Committee on Medical Research that, while the drug remained in short supply, the entire U.S. production other than that allocated to the armed forces would be assigned to one agency for clinical testing. The purpose was to ensure that maximum information would be obtained from the least amount of drug.

The CMR selected the Committee on Chemotherapeutics and Other Agents of the National Research Council to allocate supplies. Initially chaired by Dr. Perrin Long of Johns Hopkins, from June 1942 on it was chaired by Dr. Chester S. Keefer, professor of medicine at Boston University School of Medicine and director of the Robert Dawson Evans Memorial at the Massachusetts Memorial Hospitals. Other members of the committee were Dr. Francis G. Blake of Yale University School of Medicine; Dr. John S. Lockwood, University of Pennsylvania School of Medicine, Philadelphia; Dr. E. K. Marshall, Jr., Johns Hopkins University School of Medicine, Baltimore; and Dr. W. Barry Wood, Jr., of Washington University School of Medicine in St. Louis.[10,11]

PLATE 5.9.   *Chester S. Keefer, M.D., professor of medicine, Boston University School of Medicine, and director, Robert Dawson Evans Memorial, Massachusetts Memorial Hospital; World War II chairman, Committee on Chemotherapy of the National Research Council.*

PART II

Reaching for Mass Production

6

British Assessment of Penicillin's Therapeutic Activity

Great Britain had been at war since late 1939. The bombings of London and its environs had taken their toll. Although penicillin was a British discovery and its development as a useful therapeutic agent had been initiated at Oxford, the British medical profession had neither the time nor the money—much less the personnel and energy—to pursue its study further. Howard Florey had turned the problem over to the United States, and although he was impatient for results, he had no choice but to wait for action by the Americans.[1]

It was a long wait. The methods Florey had used in 1940 and early 1941 were not appropriate for prompt production of even the small amount he had requested. To develop production methods and install necessary equipment for manufacture of a drug whose therapeutic efficacy and potential market had not yet been established was more of a gamble than the U.S. pharmaceutical industry could take. Furthermore, from December 1941 on, the United States was also at war, and as in England, the demand for supplies, equipment, and personnel for essential production precluded friendly cooperation, even with Florey on a project that ultimately was so important.

But Florey and his team at Oxford continued their labors through 1941, 1942, and even into 1943. They produced small but significant amounts of the drug, which were used immediately for evaluation of its effects in humans. Finally, in March 1943, the British Medical Research Council (MRC) took over the clinical studies on penicillin that Florey had initiated. By early 1943, the British pharmaceutical industry was producing penicillin in amounts that permitted Britain to undertake an organized evaluation of its clinical efficacy, as the United States had done earlier. The need for anti-infective drugs other than the sulfonamides was great, and was forcing an evaluation of penicillin in Great Britain.

The medical profession in England had already accepted Florey's evidence that the drug was chemotherapeutically effective, but a more orderly evaluation under the auspices of the MRC seemed desirable. The British Ministry of Supply, therefore, requested that the MRC appoint a Committee on Clinical Trials of Penicillin. Dr. H. R. Dean became its chairman and Dr. Ronald V. Christie its secretary. When it came time to plan for the invasion of Europe in 1944, Christie played a major role in organizing the distribution of penicillin for care of the wounded.[2]

Background Information Available to the MRC in March 1943

Much was already known by March 1943 about the biological and pharmacological action of penicillin in animals and man. It was known that the drug was absorbed rapidly following parenteral administration and excreted rapidly by the kidneys, making frequent drug administration essential. It was known also that the drug was destroyed by gastric juice when administered orally, and that it was usually destroyed by boiling, by acids, by some heavy metals and oxidizing agents, and by enzymes produced by some bacteria. People were aware that solutions of penicillin for parenteral administration to humans had to be prepared and stored with care. It was generally believed that penicillin was usually bacteriostatic, not bactericidal, in concentrations suitable for administration to humans and that a concentration sufficient to suppress bacterial growth was probably needed at the site of infection if the drug was to be effective. Finally, it was already known that the drug could exert its antibacterial effects, at least in part, in the presence of tissue extracts, blood, and pus.[3,4] (Not until later was it shown that the action of the drug is quantitatively reduced by proteins.)

The rapid excretion of penicillin by the kidneys caused problems but was helpful, too. It meant that frequent administration (once every two to three hours) was essential; yet it meant too that a novel source of penicillin was available. Rapid excretion of the drug in urine made it possible for chemists to recover some of the excreted penicillin and for physicians to use the drug a second time (see figures 6.1, 6.2, and 6.3).

The penicillin used in Florey's initial clinical trials was impure,

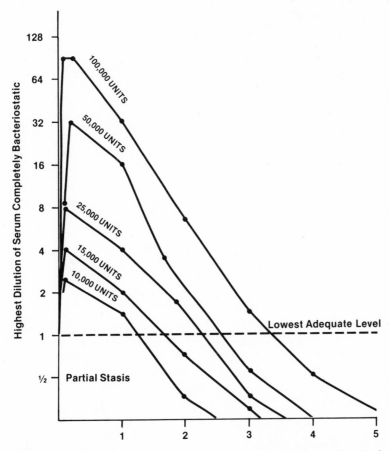

FIGURE 6.1.   Average results of bacteriostatic estimations in serum after single intramuscular injection. (From: I. W. J. McAdam, J. Duguid, and S. W. Challinor in Lancet *2 [1944]: 336.)*

unstable, and hygroscopic. Nonetheless, this very crude material was administered safely. Disappearance of infection following its administration seemed to depend on the organism's susceptibility to the drug, the use of adequate dosage, and adequate amounts of the drug reaching the infecting organisms. The necessary duration of treatment seemed to depend on the time required to eliminate the organism from lesions. When an infection was localized, Florey noted, a higher drug concentration at times could be applied to the site of infection by topical rather than by parenteral administration.[4]

Results obtained in the first 100 patients treated under the auspices of

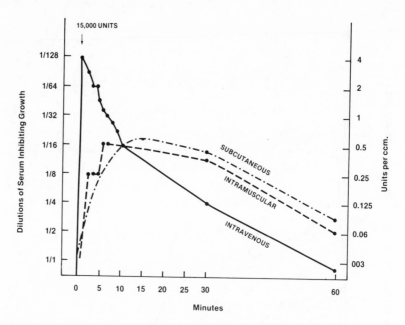

FIGURE 6.2.   Bacteriostatic power of serum after injection of 15,000 units of
penicllin by intramuscular, intravenous, and subcutaneous routes. (From:
A. Fleming, ed., Penicillin—Its Practical Application, 2d ed. [London: Butterworth
& Co., 1956], p. 72.)

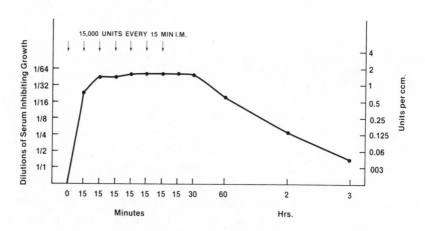

FIGURE 6.3.   Result of rapidly repeated intramuscular injection of penicillin.
(From: A. Fleming, M. Y. Young, J. Suchet, and A. J. E. Rowe in Lancet 2 [1944]:
621.)

the U.S. Office of Scientific Research and Development became available in January 1943.[5] Before March of that year, informal reports on another 100 patients began to reach England. The U.S. studies showed that parenterally administered penicillin could be highly effective in the treatment of pneumococcal pneumonia and sulfonamide-resistant staphylococcal infections. Limited observations in both the American and British armies showed, moreover, that 100,000 units of penicillin administered during a twenty-four-hour period would cure gonorrhea.[4,5]

Further, a study conducted by Dr. Mary Ethel Florey (later Lady M. Ethel Florey) in late 1942 had already established the role of topically administered penicillin. Ethel Florey had treated 172 patients with penicillin, the most important of whom were 22 patients with mastoid infections.

Following a Schwartze mastoidectomy, the wound was sewn up completely, and a fine rubber tube without lateral holes was inserted through the upper end to the base of the cavity. Penicillin solution, 250 to 500 units per cm^3, was injected 6-hourly, exudate being aspirated through the tube before each injection. Penicillin ointment was smeared on the suture line. Treatment was continued for seven days. [The patients were treated meticulously, Dr. Florey supervising the drug's administration personally, entrusting the responsibility to no one else.] Fourteen of 16 acute or subacute and 5 of 6 chronic cases healed by primary union.[3]

Eighty-nine of the 172 patients in Ethel Florey's series had had eye infections. All were due to pneumococci, staphylococci, or gonococci. Many were cured. Failures seemed to her to have been due to the persistence of relatively insusceptible infecting organisms, an associated irritation, or failure to persevere with treatment. Nine of 11 chronic wound sinuses, among the 172 patients, healed within a month after initiation of penicillin treatment, and healing was similarly observed in most of 50 patients with suppurating lesions that it seemed might lead to sepsis.[4]

Through treatment of a single patient, Alexander Fleming had also obtained important information on the effect of penicillin administered parenterally and topically.

A man, aged 52, was admitted to St. Mary's Hospital on July 7, 1942, with a history of fever since June 18th. No diagnosis was made. On July 14th, he developed signs indicating meningitis and treatment with sulfapyridine was started. Treatment was effective at first but after eight days the infection appeared to relapse. The drug was then ineffective. Sulfathiazole was administered from July 24th until August 6th at which time the infecting organisms (by then identified as streptococci) were shown to be sulfathiazole-resistant. On August 6th, treatment with penicillin, given

119

PLATE 6.1.   Lady Mary Ethel Florey with Gladys Hobby and Dr. F. Vargas Jimenez of Lima, Peru, in Buenos Aires, about 1952.

intramuscularly once every two hours was initiated. The penicillin was donated for this patient by Dr. Howard Florey. It was administered intrathecally on August 12th, 13th, 14th, 15th, and 19th, each time in a dose of 2,500 or 5,000 units. Although the course was stormy, by August 28th the patient's temperature had been normal for two weeks. He was discharged from the hospital on September 9th and remained well throughout a follow-up period of 13 months.[4,6]

By July 1942, Fleming was so accustomed to thinking of his drug as an effective chemotherapeutic agent that he said of this patient only that a man who was apparently dying of streptococcal meningitis made a complete recovery after intramuscular treatment with penicillin for eleven days and five concurrent intrathecal injections of the drug. The patient had previously received courses of sulfapyridine and sulfathiazole, and the organism had become sulfonamide resistant.

Fleming had taken care to obtain as much information as possible from this single patient, for he wished to justify the use of the large supply of penicillin he had received from Oxford. The bacteriostatic powers of the blood and cerebrospinal fluid were measured repeatedly throughout treatment. Neither the intrathecal nor the intramuscular injections caused any significant local or general reaction. A total of 1,605,000 units of penicillin were administered intramuscularly; the intrathecal injections consisted of 22,500 units of the drug. Thus, the patient had received 1,627,500 units—an enormous amount at this stage of penicillin's development.

Other information available before the Medical Research Council trials began concerned the effect of applying topical penicillin to burns and other superficial and accessible wounds and the use of topical penicillin to facilitate closure of recent soft-tissue lesions. A study was reported in the *Lancet* that showed the following:

In a group of 150 patients with "major burns" at the Royal Air Force Burns Unit, almost all of whom were infected with staphylococci, many with streptococci as well, it was observed that streptococci on the surface of a burn, unless sulfonamide-resistant, could generally be reduced in number by administration of sulfonamides, although rarely eradicated.[7] Staphylococci, on the other hand, were more difficult to control for they were only slightly susceptible to sulfathiazole and generally resistant to the other sulfonamides then available. Yet in all but three of 75 patients with burns of seven to 180 days' duration, all of whom were treated topically with penicillin, both staphylococci and streptococci were eliminated. The three failures occurred among those treated once every 48 hours. There were no failures among those treated at 24-hour intervals. Similarly, the Medical Research Council Burns Unit at the Royal Infirmary in Glasgow obtained satisfactory results—41 of 54 wounds being free of hemolytic streptococci within five days after start of treatment with an ointment containing calcium penicillin in a concentration of 120 units per gram.[8]

Thus, Fleming's single patient whom he had treated both parenterally and intrathecally, those patients treated topically by Ethel Florey and by investigators at the Royal Air Force Burns and Medical Research Council Units, and those reported from the United States (see chapter 8) had all confirmed Florey's faith in penicillin's potential, and the medical histories of the patients treated comprised part of the information available to the MRC.

As soon as supplies of the drug permitted, therefore, penicillin was shipped to the Middle East (under authorization of the director of pathology, British War Office) from Florey's laboratory for evaluation in the treatment of wounds and surgical infections in patients in the British army hospitals. Bacterial contamination occurred routinely and was not necessarily due to organisms comparable, in their susceptibility to penicillin, to those organisms causing infection in England.

The first penicillin, shipped from Oxford in July 1942, was forwarded to Cairo where Col. J. S. K. Boyd and Lt. Col. R. J. V. Pulvertaft obtained such promising results that they sought further supplies of the drug by manufacturing it themselves. No record has been found of the amount made by them, but further supplies did become available from England in November 1942 and March 1943.[9–13]

The penicillin received from Oxford probably contained no more than 40 to 50 units of active material per milligram of powder, but it was highly effective when applied locally, even to wounds that had been suppurating for long periods of time. Pulvertaft later claimed that, of all the antibacterial substances he had used, "penicillin offered the best prospect [at the time] of successfully cleaning up the numerous chronically septic conditions met with in the Base hospitals."[9–11]

Medical Research Council Clinical Trials in the Middle East, Africa, and Italy

The MRC's Committee on Clinical Trials decided in March 1943 to allocate penicillin to four designated centers for research on the systemic treatment of selected infections, to four other centers for evaluation of its effectiveness when administered topically, and most important, to the War Office for trial in wounds in the overseas army. The locations of the eight centers for systemic and topical evaluation were kept secret so as to avoid their being inundated with requests for the all-too-scarce drug.[2]

From the beginning it was agreed that, because of its scarcity, it would not be used for treatment of infections due to organisms assumed to be penicillin-susceptible, since its effectiveness in such cases had already been established. Only when new information might be obtained on the minimum effective dosage, method of administration, or other factors on which successful treatment might depend would such infections be treated with penicillin. The MRC wished primarily to explore whenever possible the desirability of penicillin treatment in conditions not previously investigated in order to confirm and extend Florey's original observations. Irrefutable evidence was needed to indicate how much could be achieved with how little drug. Since the expected supplies of penicillin had not been forthcoming from the United States and since the British pharmaceutical industry—due to the war—had been forced to direct its attention elsewhere, the number of patients who required treatment in March 1943 far exceeded the number over which supplies of penicillin could be spread.

As penicillin became available, mostly from the Imperial Chemical Industries (ICI), from Kemball, Bishop & Co., Ltd., and from Oxford, the trial of the drug in the treatment of battle casualties was extended, under the auspices of the British Medical Research Council, to North Africa. From the time of Italy's entry into World War II in June 1940, North Africa had been an active theater of military operations. Fighting had begun in September of that year and had continued almost unceasingly until May 1943. At that time, the campaigns had ended in Axis defeat, opening the way for the subsequent Allied conquest of Sicily and southern Italy.

The ultimate goal of the North African military campaign was control of the Mediterranean. The object of the penicillin trials in that area was to determine what organisms were present in long-standing septic wounds in patients seen there and to determine if well-established infections could be treated satisfactorily by the application of penicillin, used topically in most instances.

Ian Fraser, a surgeon, and Scott Thomson, a bacteriologist, received special training in the handling of penicillin at Oxford and then went to Algiers. Three weeks later in June 1943, Howard Florey, honorary consultant in pathology to the British army, followed so that he might personally supervise the study of penicillin there. He carried with him only 20 million units of the drug with which to do the job.[11] He was soon joined by Brig. Gen. Hugh Cairns, consulting neurosurgeon to the British army, who brought with him additional supplies of penicillin provided by the Therapeutic Research Corporation, an organization formed in 1941 to facilitate exchange of information within the pharmaceutical industry in Great Britain.[11,13-16]

By October 1943, the use of penicillin in North Africa had proven that superficial infections could be benefited greatly by local applications of penicillin, and it was apparent that parenteral administration of the drug, when feasible, would allow successful treatment of deep-seated infections. Initiation of treatment weeks or months after wounds were incurred, however, was less effective. Because of these findings, plans were made for earlier treatment of patients during the invasion of Sicily. Increased experience with the drug, combined with larger supplies that made parenteral administration feasible in more instances, led to outstanding results.[17]

Penicillin was still in short supply in late 1943, despite the British pharmaceutical industry's push to produce the drug in the quantity needed. Wartime demands on equipment, personnel, and funds had thwarted their efforts to produce. But by mid-1944, when the invasion of Normandy was about to begin, the Royal Army Medical Corps was in a position to plan for penicillin's parenteral use as well as its topical use as a prophylactic against infection. The study of penicillin's effect in the treatment of soft-tissue infections and infections associated with wounds, compound fractures, and burns, conducted in the Middle East, North Africa, and Italy—along with clinical reports from the United States—had provided the necessary basic information.[17-19]

Supplies of penicillin for the invasion of Normandy were made available in part by the British pharmaceutical industry. The major portion was supplied by the pharmaceutical industry in America.

7

Early Attempts at Penicillin Production in England

British investigators initially failed to isolate penicillin because suitable technology had not yet been developed. But the main problem, as indicated earlier in this book, basically was the strong conviction on the part of British investigators in the 1920s and 1930s that chemotherapy of bacterial infections was absurd and should not be contemplated. Later, because the British pharmaceutical industry was so heavily involved in wartime activities, it was unable to undertake a project of such magnitude as the development of penicillin.

Howard Florey, Ernst Chain, and their associates, however, as we have seen, recognized penicillin's potential. Their drive to realize that potential was matched by that of scientists in the United States with similar vision—Henry Dawson, Wallace Herrell, Dr. A. N. Richards, and others. These were the people who made penicillin therapy a reality.

Penicillin became a household word by 1944–45. Ever since, many have had the urge to call forth evidence suggesting that others had preceded Florey and Chain in their studies of penicillin. Unwittingly some probably did precede them, for penicillia had been used in some fermentation processes over a period of years. But there is a vast difference between obtaining a culture of a microbial strain, even cultivating it in a medium suitable for its growth and multiplication, and utilizing it knowingly for a specific purpose.

Florey and Chain's first report on penicillin stimulated some pharmaceutical companies in England to consider the possibility of producing the substance and even before its clinical effectiveness had been established, the Dyestuffs Division of Imperial Chemical Industries, Ltd., Glaxo Laboratories, Messrs. Kemball, Bishop, Ltd., and perhaps others initiated research on its production. But it was Norman Heatley who produced a sufficient amount of the drug for treatment of the first six patients; and later, in 1943 to 1944, he still produced at Oxford much of what was used in England.

Imperial Chemical Industries Attempts First Manufacture of Penicillin in England

The first report from Oxford on penicillin appeared in the August 24, 1940 issue of *The Lancet*; the second was published in the August 16, 1941 issue of the same journal. Between these two dates, on June 2, 1941, representatives of the Imperial Chemical Industries (Dyestuffs), Ltd. met with Howard Florey and Norman Heatley to discuss the possibility of the company assisting in the growth and extraction of the drug.[1-3]

Heatley reported to the Imperial Chemical Industries (ICI) representatives (Dr. A. R. Martin and Dr. C. M. Scott) all that he had done in his laboratory. He gave them his design of apparatus for the continuous extraction of penicillin and gave them details of his method of assay. The ICI representatives agreed that they would investigate at ICI the conditions needed for growth of the mold. They would thereby attempt to increase either the concentration of penicillin in the mold extract or the ease of growing the mold continuously on a large scale. Furthermore, experiments would be undertaken to evaluate drying the mold extract on a drum dryer or on a "semi-technical spray dryer" and extracting the penicillin directly onto activated charcoal from an amyl acetate extract. Heatley acquainted the representatives from ICI with his and Florey's planned trip to the United States.

Wartime pressures had convinced Florey that success could not be achieved in England, but he thought that perhaps it could be achieved in the United States. It may even have occurred to him that the more who investigated the problem, the greater the likelihood of success. Without waiting for the results of ICI's initial experiments, he and Florey took off for the United States on June 27, 1941.[4]

Howard Florey and Norman Heatley's arrival in the United States has been described previously. While they attempted to arouse interest in penicillin production in the United States, the Imperial Chemical Industries started work on the problem in England. ICI made progress, but not enough to please Florey. After nine months, he wrote, "ICI are not so good as we were when they came to see us [at Oxford]."[5] Yet Florey had left for the United States immmediately after his first meeting with ICI and at a time when he could perhaps have helped and surely could have stimulated the ICI scientists.

ICI unquestionably was the first industrial firm in England and one of the first four firms worldwide to purposefully attempt penicillin production. Within sixteen months after the meeting with Florey and Heatley, the company reported the isolation of crude penicillin as the calcium salt from a culture of *Penicillium notatum* with an estimated 25 percent overall yield. Moreover, a pilot plant had been installed at their Trafford Park facility near Manchester with an initial capacity of five 100-liter batches of fermentation liquor per week.[6,7]

Imperial Chemical Industries, Ltd., continued its research on penicil-

PLATE 7.1.   *A constant-temperature incubation room, with bottles stacked at a gentle slope to present the maximum liquid surface, at the Imperial Chemical Industries plant. (From:* Industrial Chemist, *November 1944, p. 597.)*

lin production throughout the war, growing the mold in innumerable one-quart milk bottles stacked in a slanted position that allowed maximum surface area for the mold's growth.[8] Not until about 1947 did the company consider submerged growth of the organism. In October of that year, plans were made for construction of a plant for the isolation of penicillin from 7,000 gallons per day of submerged culture liquor.[9,10]

The Therapeutic Research Corporation is Formed

Glaxo Laboratories became interested in the production of penicillin in August 1940. Unlike many of the British pharmaceutical companies, Glaxo had no long tradition of drug production in Great Britain. Rather, the firm was the successor to a New Zealand family business that had handled dried milk and other local products since 1873. The company had established itself in Britain as a manufacturer of food preparations during the early 1900s and had expanded into pharmaceuticals in 1924. It had had limited experience in the pharmaceutical field, however, when penicillin was introduced in 1940. By the end of World War II, however, as the result of its successful production of penicillin, Glaxo was firmly established in the antibiotics business.[11]

As had been the case at Imperial Chemical Industries, the impetus for Glaxo Laboratories' interest in penicillin was the paper published by Chain, Florey et al. in August 1940. Dr. F. A. Robinson, head of Glaxo's Research Division at the time, wrote on August 29 to Sir Harry Jephcott, its managing director, suggesting that Glaxo contact Howard Florey at Oxford and offer to work with him on penicillin. But Glaxo received no reply to a letter they wrote on September 26 nor to a second letter written soon after. According to W. Bryan Emery, a senior scientist on the penicillin project at Glaxo in the 1940s, "It later transpired that Florey had already contacted two other British Companies appealing for their help, had been turned down by both, and was feeling very disillusioned with British Industry."[12]

How diligently Florey investigated the possibilities of obtaining help from the British pharmaceutical companies cannot be estimated now. It is known that, in March 1941, Sir Henry Dale of the British Medical Research Council and a representative from Burroughs Wellcome Company, visited Florey to inquire about penicillin. Representatives of Boots Pure Drug Company also visited with him in June 1941 and learned of the techniques then in use at Oxford and of the results being obtained.[5] The importance of large-scale production was stressed at each meeting, but it must be recalled that in March 1941 Florey had virtually no evidence that penicillin would be therapeutically effective and in June, he had only his experience with a few patients to report. Nothing came of the two meetings.

In the meantime, Florey's team was encountering difficulties. Its methods for the extraction of penicillin were impractical for large-scale production. The team was making only about 2 grams of penicillin per week and that was only about 1 to 2 percent pure.[9] At about this time, Prof. Harold Raistrick at the London School of Hygiene and Tropical Medicine became a consultant to Glaxo Laboratories. Raistrick had had experience with penicillin—in 1933 he had studied methods for its isolation.[13] Glaxo received nine strains of penicillium from Raistrick, one of which, B.592 (Fleming's strain NIMR 1929, obtained from the National Institute of Medical Research in London) proved consistently best. This strain was therefore chosen for the company's work on penicillin production. The methods devised by the Oxford team were used at first, and in general the Oxford results were confirmed.

On December 2, 1941, representatives of Burroughs Wellcome & Company, Boots Pure Drug Company, British Drug Houses, and Glaxo Laboratories met to discuss penicillin production. It appeared from work that was described at the meeting that there were at least two different forms of penicillin—penicillin A (later identified as notatin)[14] which was under investigation at Boots, and penicillin F (2-pentenyl penicillin) which was being studied by Florey, Burroughs Wellcome & Company, and Glaxo Laboratories.[15] It was decided that they would continue work on both and would accumulate larger quantities of each for identification studies.

Thus, by late 1941, at least five firms in England were actively engaged in studies on penicillin production—this despite wartime restrictions on personnel, equipment, and supplies.[12,16]

In January 1942, the Therapeutic Research Corporation of Great Britain, Ltd. (TRC) was formed by the mutual agreement of Boots, the British Drug Houses, Ltd., Burroughs Wellcome and Company (the Wellcome Foundation, Ltd.), Glaxo Laboratories, Ltd., and May and Baker, Ltd. (a chemical firm known particularly for its production of sulfonamides, notably sulfapyridine, also known as M & B 693). The first meeting of the corporation was held on January 16.[12,16]

The formation of the TRC was the British pharmaceutical industry's first attempt to organize their research and pool results on specified

129

subjects (see note 16, chapter 6). Although it was not intended, the study of penicillin almost immediately assumed first priority among the corporation's activities; major contributions related to its production and to understanding of its chemical nature were made by the member companies. Later, in 1949, the National Research Development Corporation (NRDC) was established under the British Industries Act of 1948 in order (1) "to secure when the public interest so requires the development or exploitation of inventions resulting from public research and any other invention ... not being sufficiently developed or exploited," and (2) "to acquire and grant rights in connection with inventions (including new techniques) gratuitously or for payment." The NRDC was Great Britain's reply—admittedly somewhat late—to U.S. investigators obtaining patents on processes related to penicillin production. It was its mechanism for protecting itself against any future patents.[16-18]

In November 1942, the Imperial Chemical Industries joined the TRC, and at the request of the Ministry of Supply, a committee was formed to coordinate work on the production and purification of penicillin. (Up to this time, the TRC had been concerned primarily with research.) Alexander Fleming and Howard Florey were invited to join the committee, together with a representative of the National Institute of Medical Research.[12,16]

Glaxo Laboratories Increases Penicillin Output

By December 1942, Glaxo had a small production unit at Greenford which produced about 70 liters of broth per week with a potency of approximately 150 units per milliliter. The efficiency of the extraction process was approximately 30 percent and the potency of the resultant solids was about 150 to 200 units per milligram of calcium salt.[12] In February of the next year, Glaxo started operating a second factory at Aylesbury. The methods used were not unlike those developed at Oxford, but in the spring of 1943 the firm learned of the United States' success with corn steep liquor, and in September it converted entirely to the use of a lactose–corn steep liquor medium. At the same time, it began to use the American penicillium strain obtained from Dr. Robert Coghill of the Northern Regional Research Laboratories in Peoria (strain NRRL 1249.B21).[12]

The production of penicillin from the two Glaxo plants in 1943 was 2,570 million units, part of the United Kingdom's total output of 3,500 million units. Later, in February 1944, Glaxo opened a third factory at

PLATE 7.2.   *Germinators for feeding 300 gallons of inoculum to 10,000-gallon fermentors at Glaxo Laboratories. (From: A. Fleming, ed.,* Penicillin—Its Practical Application, *2d ed. [London: Butterworth & Co., 1950], p. 41.)*

Watford; the company's total production from January to June 1944 was about 7,500 million units as final dried salt. In January 1945 a fourth plant was opened at Stratford, which produced in the first nine months of the year 35 billion units.[12]

Up to D-Day, June 6, 1944, production from the Glaxo factories comprised about 80 percent of the country's total output, and at no time during the war was it below 50 percent.[12] Most of the penicillin produced at this time was sent to the Medical Research Council Clinical Trials Committee[19] for evaluation of its therapeutic potential or to Oxford University or the Imperial College in London for chemical studies. Distribution of all penicillin produced in England prior to 1946 was controlled by the British government.

131

Kemball, Bishop & Company Takes a Step Toward Large-Scale Production

Kemball, Bishop & Company, Ltd. was by no means among the first to undertake research on penicillin production. For many years, Kemball, Bishop had made citric acid for the European market; in 1936, according to an agreement between Kemball, Bishop & Co., Ltd. and Chas. Pfizer & Co., Inc., representatives from the Pfizer company completed construction of a surface fermentation citric acid plant in England. Since 1923, Pfizer had manufactured citric acid by surface fermentation with a strain of *Aspergillus niger*. Thus Kemball, Bishop's interest in fermentation processes was well known when penicillin came along. When Howard Florey failed to obtain from other sources the supplies of penicillin he wanted, he turned to Kemball, Bishop for assistance.[20-22]

The firm started work on penicillin on March 5, 1942, with Prof. Robert Robinson of Oxford and Prof. Harold Raistrick as advisers. On

PLATE 7.3. Aerial view of the Crown Chemical Works, which was the site of the Kemball, Bishop factory. Arrow indicates Factory 14 where penicillin production was carried out under high-security conditions, 1942.

that day, Florey's culture of *Penicillium notatum* was subcultured for the first time at Bromley-on-Bow on the outskirts of London where the Kemball-Bishop plant was located. L. M. Miall and V. J. Ward were responsible for the fermentation studies and for the assay of penicillin solutions. The mold was grown at first in flasks, but almost immediately Kemball, Bishop began conversion to a tray fermentation process similar to that used for the company's production of citric acid. This required that filtered air be installed in order to avoid contamination of the broth and required also an air-cooling system to give a fermentation temperature of 24°C. On March 25, 1942, only three weeks after they had started work on the drug, the first trays were seeded for production of penicillin in England on a scale larger than that possible in glass bottles. The yields in these first trays were up to 7 Oxford units per milliliter.[22]

From then on, Kemball, Bishop made steady progress in the fermentation of penicillin. However, assemblage of the necessary extraction apparatus proved difficult under the conditions existing in the East End of

SOME OF THE DESTRUCTION AROUND THE CROWN CHEMICAL WORKS.

PLATE 7.4. Photograph taken after the blitz of East London, 1942. The Crown Chemical Works remained standing but surrounded by devastation.

London in wartime. So in the autumn of 1942, they started sending penicillin broth in milk churns by lorry to Oxford for extraction in the equipment at the Sir William Dunn School of Pathology (see plates 7.3 and 7.4).

On October 28, 1942, Florey wrote to thank Kemball, Bishop for "150 gallons of brew, at 9 units per ml., a total of 6.1 million units...a handsome supply." Florey considered Kemball, Bishop "quite the most likely to produce penicillin in quantity in the near future. Others have 'plans' but translation of these 'plans' into action is a different matter."[21]

For some years there were six 20-gallon stainless steel fermentors at Kemball, Bishop that were often referred to as "the old penicillin fermentors." The fermentors were designed as such, incorporating many precautions against ingress of contaminating bacteria. But they were never used as such; they played no part in the development of penicillin.

Kemball, Bishop was convinced that its technique for growing the mold by surface culture on trays was so much better than the methods used in the bottling plants of other British producers that they did not investigate production by submerged culture. Unlike Pfizer, Kemball, Bishop had had no previous experience with submerged growth of molds. They had no staff available to research the process, for they were fully employed with production. Their faith in surface growth in pans was so great that they failed to heed communications from John L. Smith, then vice-president of Pfizer, who on several occasions wrote detailing the company's strong interest in, and implied success with, submerged fermentation for production of penicillin.

The "old penicillin fermentors" were never used for their original purpose. They performed a notable role, however, in that they were the first stainless steel deep fermentors to be used by Kemball, Bishop for citric acid production. They were used also in the development of gluconic acid and some other organic acids.[20-25]

Submerged Penicillin Fermentation in England

At the beginning of 1942, the Imperial Chemical Industries received information from America indicating that in the United States the production of penicillin by submerged fermentation was becoming a reality. A recommendation was made therefore to the British Ministry of Supply that a mission be sent to the United States to exchange information on penicillin research and production. Dr. W. R. Boon of ICI and Dr. W. B.

134

Hawes of the Wellcome Foundation, Ltd. were designated to carry out this mission.

In March to April 1943, they visited on behalf of the British Ministry of Supply fourteen commerical laboratories in the U.S.A. (i.e., Abbott Laboratories, the Chester County Mushroom Laboratories, Hoffman LaRoche & Co., Inc., Lederle Laboratories, Eli Lilly & Co., Inc., Merck & Co., Inc., Parke Davis & Co., Inc., Chas. Pfizer & Co., Inc., the Schering Corporation, E. R. Squibb & Sons, Frederick Stearns & Co., and the Winthrop Chemical Company) where penicillin production was under investigation and met with Dr. Robert Coghill at the NRRL, Dr. Chester Keefer, Dr. A. N. Richards, and Dr. Selman Waksman at the New Jersey Agricultural Experiment Station (Rutgers University) where he had long been studying antibiotic substances in nature. Of the fourteen firms visited, only Merck & Co., E. R. Squibb & Sons, Pfizer, and the Chester County Mushroom Laboratories were at that time producing material routinely for clinical use; Abbott Laboratories and Eli Lilly & Company were producing enough penicillin for their own privately organized clinical trials. Boon and Hawes returned to England with full details on what they had seen and learned of America's penicillin program.[26]

Boon made a second trip in February to March 1944, this time accompanied by Sir Harry Jephcott, managing director of Glaxo Laboratories. He found many changes had taken place in the one-year period. The War Production Board had assumed full responsibility for penicillin production, and under its aegis, the United States was pushing toward a goal of more than 380 billion units of penicillin monthly.[27] Surface fermentation was still widely used, although more and more emphasis was being placed on submerged growth of the mold. Boon and Jephcott's report to the Ministry of Supply, written upon their return to England, contained no mention of Pfizer's deep tank fermentation plant opened less than a month later.[10, 26-29]

The year before, Glaxo had first carried out experiments on deep culture fermentation in January 1943 in four-gallon vessels, but they had obtained very low titers. Because of the need to concentrate on maximum production coupled with a shortage of staff, they had made little progress on the new process. Sir Harry Jephcott, therefore, while on his visit to the United States, made an agreement with two American companies in April 1944 to exchange information. Jephcott after his visit was convinced that deep culture fermentation was the method of the future.

Unfortunately, Boon was not as convinced, possibly because the ICI engineers had devised, on paper, a process for continuous surface fermentation. Interestingly, at the close of Boon's first visit to the States with Hawes, both men had expressed the opinion that the deep fermentation process offered the only possibility of obtaining adequate supplies of penicillin within "the reasonably near future."[30] Following a design based on the continuous fermentation process of their own devising, ICI built an elaborate surface culture plant that consisted of large horizontal cylindrical autoclaves containing tiers of trays fed by a cascade system. Although the process worked, it was costly and the amount of penicillin produced was trivial.[28-31]

Boon's attitude made it difficult—indeed impossible—for Jephcott to get the support of the Ministry of Supply, and he was forced to cancel his agreement with the Americans. Admittedly, penicillin was needed, surface fermentation assured a good product if not large quantities of it, and even the U.S. War Production Board required for a time that some manufacturing firms use surface fermentation only.

Later in 1944, however, the Ministry of Supply finally arranged for the Commercial Solvents Corporation (a U.S. firm that had become involved in penicillin production in 1943) to install a deep culture plant at Speke, near Liverpool, and requested that Glaxo build the submerged fermentation plant that Jephcott had proposed earlier in the year. Glaxo accomplished this assignment with the cooperation of Merck & Company and E. R. Squibb & Sons of the United States. A plant for the production of penicillin on a 5,000-gallon scale went into operation at Barnard Castle in County Durham in January 1946. It was so successful that it produced far more in the first nine months of 1946 than had been produced by all three of the Glaxo surface culture plants in the whole of 1945. In addition, the penicillin from the Barnard Castle plant was about 80 percent pure. The Glaxo surface culture plants were closed later in 1946.

While all this was transpiring at Glaxo in 1944 to 1945, the British Ministry made plans for still another deep tank fermentation plant at Speke, although the available plants for surface fermentation were kept in operation.

The firms engaged in surface production of penicillin in England were all pharmaceutical companies with no previous experience in fermentation. In mid-May 1944, representatives of the British Ministry of Health and Sir Graham Hayman, chairman of the Distillers Company, finally met to discuss penicillin production. Dr. John Hastings of the Distillers Company was invited by Sir Graham to attend as his technical adviser. Until 1940 Hastings had been in charge of an acetone-butanol fermentation plant which had been started in 1935 on behalf of the Commercial Solvents Corporation. This plant had been sold to Distillers before World War II.[30]

At the mid-May meeting between Sir Graham and representatives of the Ministry of Health, the ministry proposed that Distillers turn over one or more fermentation plants so that selected pharmaceutical companies could engage in deep fermentation. Sir Graham refused, knowing that Distillers had almost all the existing expertise in the country. Instead, he proposed that Distillers itself do the job, with cooperation from the Commercial Solvents Corporation if the ministry insisted.

The Commercial Solvents Corporation (CSC) in the United States thus became a joint agent on the project and gave all its information about the process to the Distillers Corporation. Ten days later, on May 28, 1944, Distillers sent a team to visit CSC. Making the trip were Hastings, G. W. Daniels (chief engineer), Charles Grover (chief research bacteriologist), and later J. Everett (general works manager). They flew from Croydon in a Dakota, blacked out so they could not see the forces along the south coast of England gathering for the invasion of Normandy. They landed first at Shannon and then flew in a U.S. Navy Sikorski flying boat to New York. There they remained until August when the British and the U.S. governments finally agreed that the team could visit the CSC plant at Terre Haute, Indiana. This plant at that time was producing penicillin steadily by submerged growth only.

Hastings and his associates returned to Great Britain on September 17 with technical information, cultures of the strain of *P. notatum* used by CSC and 21 rabbits and 120 mice for test purposes.[30] By the time they reached England, a site at Speke had been selected for construction of the plant. Distillers already had other fermentation plants near Speke as well as available materials and technical staff. Liverpool's location on a tidal estuary with thirty-six-foot tides would allow the fermentation effluent to be discharged directly into the estuary without treatment.

The CSC plant in the United States consisted of ten 10,000-gallon fermentors. Their average yield was 100 Oxford units per milliliter. The Speke plant was twice this size. Construction began in December 1944, the plant was completed by December 1945, and the first full-scale test run was completed on December 12 of that year. The first approved batches of

PLATE 7.5.   *Sterilizers for air entering fermentors in deep culture factory at Distillers Company. (From: A. Fleming, ed.,* Penicillin—Its Practical Application, *2d ed. [London: Butterworth & Co., 1950], p. 39.)*

penicillin from the Speke plant were shipped on December 28.[30] The plant immediately produced the anticipated quantities of the drug, with manufacturing costs reduced to 5 percent of those of surface culture production. Bottle plants throughout England were soon closed, for their joint output was equivalent to only a small fraction of that obtainable by submerged fermentation at the Speke plant.

In building the Speke plant, CSC and Distillers had operated as joint agents of the Ministry of Supply with only an informal agreement on arrangements. After the plant had begun to function successfully, it became clear that continued research and development would be necessary. The British Ministry of Supply refused to finance this, and CSC withdrew from the agreement. Distillers Corporation, as sole agent thereafter, continued operation of the plant until the British government sold the plant to the company for commercial operation in 1947.

At the Speke plant, from 1945 on, production increased constantly. Mutant cultures, some first produced in the United States and others produced later in Distillers' own laboratories, were used. New types of extraction equipment gave high yields and produced concentrates from which crystalline penicillin could be obtained easily. The use of precursors in the fermentation, moreover, led to total production as penicillin G (benzyl penicillin).

The large and elegant surface culture plant that ICI had constructed earlier became outmoded as Glaxo moved into deep tank fermentation. But ICI had an excellent purification plant and an underused extraction process, so the firm made an important contribution by extracting penicillin from the fermentation liquors produced by others. Meanwhile ICI closed down its surface production and converted to vertical cylindrical closed fermentors.

Later, Boots of Nottingham, which had previously operated one of the largest surface culture plants in England, also received assistance from ICI in the extraction of penicillin from its fermentation liquors while Boots established its own deep fermentation plant.

By November 1945, large-scale production units had been completed or were under construction in England at Allen & Hanbury's in Ware; Boots Pure Drug Company in Nottingham; Distillers Company at Speke; Glaxo Laboratories at Watford, Stratford, and Barnard Castle; Imperial Chemical Industries, Ltd. at Trafford Park near Manchester; Kemball, Bishop, Ltd. in London; the Royal Navy Medical School in Clevedon, Somerset; and the Wellcome Foundation in Belmont.[11]

From a small beginning in Howard Florey's laboratories in Oxford in 1939, production of penicillin in England had gradually increased to an estimated 27 or more million units per week by March 1943 and, as interest in deep tank fermentation grew, to an estimated 25 to 30 billion units per week by 1945–46.[32-36]

8

America Evaluates Penicillin's Potential: Therapeutic and Economic Considerations

On August 5, 1943, the *Journal of Commerce* reported that, although the foremost authorities in the United States had been enlisted to increase the production of penicillin, the civilian population would not benefit from its therapeutic properties "for years to come."[1] The author of this report had failed to recognize the ingenuity of the U.S. pharmaceutical industry and the riskiness of trying to predict the future.

Two years earlier—in June 1941, almost two years before the British Medical Research Council appointed its Committee on Clinical Trials of Penicillin—the Committee on Medical Research (CMR) had been established in the United States within the Office of Scientific Research and Development (OSRD). Dr. Vannevar Bush, director of the Carnegie Institution in Washington, D.C., was director of the OSRD, which quickly assumed major responsibility for the entire United States' program of civilian scientific research and development, including, among many other things, all aspects of military medicine.[2-4]

Dr. Alfred N. Richards, vice-president and professor of pharmacology at the University of Pennsylvania in Philadelphia became chairman of the CMR. The advisory services of the Division of Medical Sciences of the National Research Council (NRC) were enlisted, and the chairman of the Division of Medical Sciences of the NRC was elected vice-chairman of the CMR. By March 1943, the two groups were already working in unison.

The goals of the CMR as they related to penicillin differed from those of the British Committee on Clinical Trials of Penicillin. In the United States, the goal was to establish the breadth of penicillin's therapeutic

141

potential and the wisdom (from an economic standpoint) of embarking on its large-scale manufacture. In Great Britain, it was to see how much could be accomplished therapeutically with how little drug.[5,6]

Although penicillin was still scarce in the United States, it was available in far greater quantities than in England. Greater emphasis could be placed on its effectiveness when administered parenterally. From March 1942 until January 1943, the entire supply of penicillin that was produced in the United States for clinical testing was turned over to the CMR for use by the OSRD and the armed forces at no cost to the government, physicians, or the patients who were treated with it. Beginning in January 1943 and continuing until December 31, 1945, the OSRD purchased the penicillin through a contract with the Massachusetts Memorial Hospitals. The contract stipulated that the drug would be distributed to other hospitals through their accredited "responsible investigators" who in turn would furnish to Dr. Chester Keefer in Boston information concerning the results of administration of the drug. The OSRD paid the industry $200 per 1 million units of penicillin—an amount set by the industry itself but far below its manufacturing cost. By 1945, the cost of production had decreased so much that the amount paid dropped to $6 per 1 million units.[1-7]

In 1942 and during the first six months of 1943, the supply of penicillin in the United States was so limited that only twenty-two accredited investigators were selected to receive it. These were the most prestigious and knowledgeable infectious disease specialists in the country. With the amounts of drug available to them, they attempted to evaluate its effectiveness in the treatment of staphylococcal infections and, to a more limited extent, infections caused by streptococci or pneumococci that had failed to respond to sulfonamide therapy.

On January 1, 1943, Keefer reported the results obtained in the first 100 patients treated with penicillin distributed by the OSRD. By March, Keefer was able to report on the treatment of 200 patients, and by August, on 500 who had received penicillin therapy. News of the results of treatment spread by word of mouth, and informal and government reports.[8,9]

All penicillin produced in the United States was placed under allocation by the War Production Board (WPB) on July 16. The WPB allocation order assigned all penicillin produced in the United States to the U.S. Army, Navy, and Public Health Service; that portion meant for research purposes went to the OSRD. Production by then was great enough to

PLATE 8.1.   Wallace E. Herrell, M.D., of the Mayo Foundation and Clinic, staff
member who was one of the first twenty-two investigators accredited by the OSRD
for clinical studies of penicillin. (Courtesy of the Mayo Foundation.)

permit its distribution to all physicians throughout the country who were
concerned with the treatment of infections in those disease categories of
interest to the Committee on Medical Research.[4,10,11]

By May 1944, the needs of the armed forces and our allies were so well
met and the amount of penicillin being produced had so increased that the
WPB was able to allow limited sale of the drug through hospitals. The
Massachusetts Memorial Hospitals program under Dr. Keefer then was
limited to special investigators and to special studies requiring long-term
support. The program was terminated at the end of 1945 with data on
10,838 patients available for analysis. The results of this analysis were

143

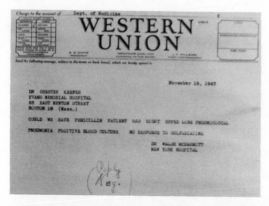

PLATE 8.2.   *Request for release of penicillin by the Office of Scientific Research and Development. Prior to December 1943, even the most respected clinicians had difficulty getting the drug for their patients.*

published in 1948, but interim reports made the data usable long before that.[12]

During 1943, 21 billion units of penicillin had been produced; in the next year, production rose to more than 1,600 billion units. By December 1944, more than 2,700 hospitals were receiving monthly supplies of the drug and regularly placing orders for their needs, although not all controls had been removed. On December 12, 1944, Fred J. Stock, chief, Drugs and Cosmetics Branch, Chemicals Bureau, War Production Board, reported to the American Pharmaceutical Manufacturers Association: "Production has met military requirements and has permitted increased quantities to civilians. Peak production has not yet been reached and the upward trend of production will continue for several months." Production did continue to increase in the months ahead.[10]

Still larger amounts per month were being produced by early 1945, and on March 15, the WPB released penicillin for sale through normal trade channels in vials containing 100,000 units each for human parenteral administration. By then, limited quantities were also being allocated to the Foreign Economic Administration for export to areas where distribution controls were comparable to those in use in the United States.[11]

Thus, the breadth of penicillin's potential had been established by the end of 1943, and the U.S. pharmaceutical industry, hesitant at first to attempt large-scale production of the drug, had achieved the impossible long before the Allied Forces' invasion of Normandy on D-Day.

Penicillin Released by the OSRD—The First Patient

The first patient to receive penicillin allocated through the OSRD was Mrs. Ogden Miller, wife of the athletic director at Yale University.[13-15] Mrs. Miller had developed a streptococcal septicemia after a miscarriage, and after four weeks of 103° to 106° temperatures, she was given a first dose of penicillin:

It [the penicillin] arrived air-mail [from Merck & Co. in Rahway, New Jersey] Saturday morning (March 14, 1942) and a small trial dose was given at 3:30 Saturday afternoon. This was tolerated well so they then gave her larger doses every four hours. By 9 A.M. Sunday her temperature was normal for the first time in four weeks and has stayed normal until this writing (noon Monday). She has eaten several enormous meals—also for the first time in four weeks. It really looks as though Florey had made a ten-strike of the first water.

One week later, Mrs. Miller's response to the penicillin was described more fully:

This week the hospital has been very excited because of the extraordinary results which have followed...[the] clinical trial of Howard Florey's penicillin.

I mentioned last week the case of Mrs. Ogden Miller, who for four weeks had been going downhill with what appeared, on the basis of all previous experience, to be a fatal hemolytic streptococcus septicemia. She had had a temperature [steadily] ranging from 103° to 106.5°...for four weeks despite liberal administration of the sulfa drugs and had 50–100 or more bacteria per cc [i.e., cubic centimeter] of blood. On Saturday the 14th, when the first dose of penicillin was given, her bacterial count was still well over 50 and [her] temperature [was still] 105°. Her temperature had returned to normal by 4 A.M. Sunday, she had 1 bacterium per cc that day and Monday, and thereafter three completely sterile blood cultures on successive days.... Monday, March 23rd...[the] blood culture [was] still sterile. It is still too soon to say that she is cured but the response has been most dramatic.[15]

The Treatment of Acute Infections

Early in the OSRD study of penicillin, and indeed in some instances even before distribution of penicillin through the OSRD, other patients with pneumococcal and hemolytic streptococcal infections received small amounts of the drug. Pneumococci and hemolytic streptococci were among those organisms that Fleming had shown in 1929[16] to be most susceptible to the growth-inhibitory action of penicillin, and it was not difficult to confirm this. Only one to two one-hundredths of a unit of penicillin were required for in vitro inhibition of the growth of pneumococci, and observations soon established that the substance was similarly active against hemolytic streptococci.

Pneumococci and hemolytic streptococci were among the most prevalent and most virulent disease-producing microorganisms in the early 1940s. Of particular concern was the high incidence of pneumonia, an inflammatory condition characterized by the formation of an exudate in the interstitial and cellular portions of the lung and a resultant consolidation of lung tissue. More than 95 percent of the cases of pneumonia at the time were caused by pneumococci (*Diplococcus pneumoniae*, now designated *Streptococcus pneumoniae*), 2 to 3 percent by hemolytic streptococci (*Streptococcus pyogenes*), and another 1 percent by strains of *Staphylococcus aureus*. Occasional cases were due to other organisms, for example, Friedlander's bacillus (now known as *Klebsiella pneumoniae*)—a Gram-negative rod-shaped organism characterized by a heavy slimy capsule, widely distributed in nature and commonly found in the intestinal tract of animals and man—and *Hemophilus influenzae*—a small Gram-negative coccobacillary strict parasite requiring special factors for growth on artificial culture media, found normally in the nasopharynx and once thought (erroneously) to be the etiological agent of pandemic influenza.

What the actual incidence of pneumonia was then is uncertain, for reporting techniques were not good and the need for records was not widely recognized. The pneumonia mortality rate for the period, however, reveals that despite a continuing downward trend from 1918 on, and despite the availability of specific horse antiserum from 1933 to 1934 on, rabbit antiserum from 1935 on, and the sulfonamides from 1937 to 1938 on, the pneumonia death rate in 1940 approximated 50 per 100,000 population.[17]

Even without direct complications, pneumococcal pneumonia was very often fatal. With complications, as, for example, empyema, pericarditis, endocarditis, meningitis, or bacteremia—none unusual in patients with pneumococcal pneumonia—fatalities were even more frequent. Penicillin abolished the fatalities that were due to pneumonia itself. It only decreased the frequency of death due to complications.[18]

Pneumococcal and Streptococcal Infections. Two patients with acute pneumococcal endocarditis came to the attention of Henry Dawson in 1942.[19, 20] Both were septicemic—that is, the infecting organisms had spilled into the bloodstream. Both were refractory to sulfonamide therapy. No remedy was available, except penicillin which was still scarce and still untested. The two patients were treated with penicillin as intensively as was then possible, and although both ultimately succumbed to their infection, there was dramatic initial improvement with temporary sterilization of the blood. Although the fulminating pneumococcal infections were not brought under control, temporary sterilization of the blood of these patients represented a remarkable achievement.

Soon after, Dr. William Tillett and his associates at the New York University College of Medicine and Bellevue Hospital in New York City[21, 22] reported such striking results in the treatment of pneumococcal infection that no one could question them. During the seasons of 1942–43 and 1943–44, 110 patients with pneumococcal pneumonia were treated with penicillin. The total mortality (7 cases) was only 6.3 percent. Bacteremia, which occurred in 40 of the patients, was successfully eliminated in all instances. In 3 instances, infections caused by sulfonamide-resistant strains of pneumococci were treated effectively. In addition, 20 of 21 patients with pneumococcal empyema responded well when penicillin was administered intrapleurally.

Staphylococcal Infections. Forty-seven of the first 100 patients treated under the auspices of the OSRD suffered from staphylococcal infections.[8,9,20] Among 32 of these patients—all with bacteremia—there were 11 deaths. If one excludes 6 with endocarditis and those others who received inadequate amounts of the drug, the fatality rate in this group of bacteremic patients was only 12.5 percent. In contrast, prior to the introduction of penicillin, the usual gross fatality rate in this type of infection in the early 1940s approximated 85 percent without sulfonamide and 60 percent with sulfonamide treatment.

The treatment of local staphylococcal infections was just as remarkable. Thirteen of 15 patients with chronic osteomyelitis improved to the extent that the sinuses were sterilized and the lesions healed. In 2 patients with mixed infections of long standing, the results were less satisfactory.

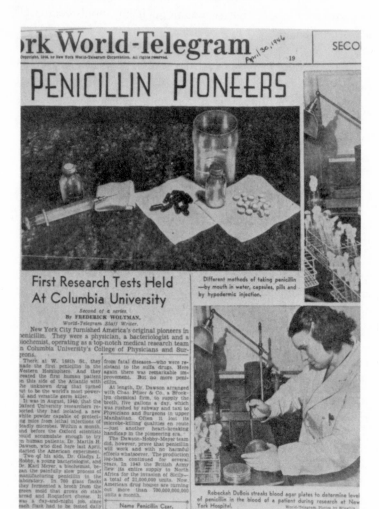

PENICILLIN PIONEERS

Different methods of taking penicillin
—by mouth in water, capsules, pills and
by hypodermic injection,

First Research Tests Held At Columbia University

Second of a series.
By FREDERICK WOLTMAN,
World-Telegram Staff Writer.

New York City furnished America's original pioneers in penicillin. They were a physician, a bacteriologist and a biochemist, operating as a top-notch medical research team in Columbia University's College of Physicians and Surgeons.

There, at W. 168th St., they made the first penicillin in the Western Hemisphere. And they treated the first human patient on this side of the Atlantic with the unknown drug that turned out to be the world's most powerful and versatile germ killer.

It was in August, 1940, that the Oxford University researchers reported they had isolated a new white powder capable of protecting mice from lethal injections of deadly microbes. Within a month, and before the Oxford scientists could accumulate enough to try on human patients, Dr. Martin H. Dawson, who died here last April, started the American experiment.

Two of his aids, Dr. Gladys L. Hobby, a young bacteriologist, and Dr. Karl Meyer a biochemist, began the painfully slow process of manufacturing penicillin in the laboratory. In 700 glass flasks they fermented a broth from the green mold that grows on stale bread and Roquefort cheese. It was a day-and-night job, since each flask had to be tested daily

from fatal diseases—who were resistant to the sulfa drugs. Here again there was remarkable improvement. But no more penicillin.

At length, Dr. Dawson arranged with Chas. Pfizer & Co., a Brooklyn chemical firm, to supply the broth, five gallons a day, which was rushed by subway and taxi to Physicians and Surgeons in upper Manhattan. Often it lost its microbe-killing qualities en route —just another heart-breaking handicap in the pioneering era.

The Dawson-Hobby-Meyer team did, however, prove that penicillin will work and with no harmful effects whatsoever. The production log-jam continued for several years. In 1943 the British Army flew its entire supply to North Africa for the invasion of Sicily— a total of 21,000,000 units. Now American drug houses are turning out more than 780,000,000,000 units a month.

Name Penicillin Czar.

Rebeckah DuBois streaks blood agar plates to determine level of penicillin in the blood of a patient during research at New York Hospital.

World-Telegram Photos by Bitzilla

PLATE 8.3. "Penicillin Pioneers—First Tests Held at Columbia University." (Reproduced from: the New York World-Telegram, April 30, 1946.)

Keefer described the results in the treatment of osteomyelitis as "impressive." The same was true for other types of localized staphylococcal infections. In the treatment of empyema and burns, local application of penicillin was highly effective, while the parenteral route of administration was the method of choice for the treatment of carbuncles and deep-seated infections.

It was obvious that penicillin would be of great value in the treatment of staphylococcal and streptococcal infections, both exceedingly common in war wounds and burns.

Surgical and Wound Infections. On April 1, 1943, the Office of the Surgeon General, U.S. Army, established a unit for penicillin therapy at the Bushnell General Hospital in Brigham City, Utah, and on June 3 of the same year, a second unit was established at the Halloran General Hospital on Staten Island, New York. These two units functioned as training centers for selected medical officers who subsequently participated in the evaluation of penicillin's effectiveness in the treatment of surgical infections. The project received the full cooperation of the Office of Scientific Research and Development.[23-25]

Sufficient drug was available to permit its administration parenterally as well as locally, and it was possible to study both its therapeutic and its pharmacological effects. An attempt was made to prevent cross infection and contamination of wounds. The study included evaluation of: (1) methods of penicillin administration, optimum dosage, and resultant side effects; (2) the drug's effectiveness in the treatment of acute pyogenic infections; and (3) its role in the treatment of chronically septic compound fractures. In a summary report of the study, written at its conclusion, evidence of the remarkable antimicrobial power of the drug was presented and it was concluded that "the importance of the study lies in the demonstration that penicillin permits active surgical intervention almost immediately.... The bacteria present in the wounds however are variously susceptible to penicillin and are important limiting factors in the choice of operative procedure in a given case."[25] The results of the study were impressive and gave strong support for the belief that penicillin must become commercially available. Many believe that it was the most significant clinical trial conducted with penicillin in the early 1940s.

Gonococcal Infections. Gonococci, the etiological agents of gonorrhea, are Gram-negative microorganisms that appear in the form of diplococci, usually paired and characterized by the fact that their adjacent sides are flat or concave, giving the pairs a biscuit-shaped appearance. In

149

stained smears of exudates from patients, they usually are contained within polymorphonuclear leucocytes. The organisms are obligate aerobes and are remarkably fastidious in their growth requirements when cultivated artificially in vitro. The portal of entry into humans is usually by way of the genitourinary tract, both in males and females—the primary infection occurring in the genitourinary tract although manifestations of the infection may be apparent in many parts of the body. Infection is almost always the result of sexual contact.[26]

Gonococci, properly known as *Neisseria gonorrhoeae*, are parasites of man. There is no evidence that they ever spontaneously cause disease in animals, and experimental gonococcal infection in animals is produced only with difficulty. For these reasons the evaluation of penicillin's action against gonorrhea was greatly handicapped in the early 1940s. Yet, it was clear from Alexander Fleming's initial report on penicillin that gonococci were among the most penicillin-susceptible of the organisms he tested.[16]

Attempts to establish persisting gonococcal infection in the vaginal mucosa of immature female mice that were highly susceptible to many other infections and to establish gonococcal conjunctivitis in the eyes of rabbits were unsuccessful. Success would have led to easy evaluation of the action of the many sulfonamide compounds then being developed, as well as of penicillin. Since experimental gonococcal infections that were suitable for drug evaluation apparently could not be produced by methods then known, Dr. John Mahoney and his associates of the U.S. Public Health Service, in collaboration with the Federal Bureau of Prisons, undertook a drug evaluation program in volunteer human subjects at the federal penitentiary in Terre Haute, Indiana. Their goal was to determine if prophylaxis against gonorrhea could be achieved. Results of the study in volunteers, however, showed only that it was not possible to produce experimental infection with any degree of regularity—even in humans— either by inoculation of test tube–propagated gonococci or by direct transfer of pus from patients with natural or experimental gonorrhea. The disease became apparent in only about one-half of the volunteers, and no information on the ability of either penicillin or the sulfonamides to prevent gonorrhea was obtained.[27]

Despite all this, the high incidence of gonorrhea and the remarkable susceptibility of the infecting organism to the action of penicillin led to its attempted use in treatment of the disease. At times, patients were admitted to the hospital, the diagnosis was confirmed bacteriologically, and the patient was treated, cured, and discharged before hospital authorities knew of the admission. By 1943 to 1944 many reports were forthcoming, indicating its powerful activity against gonorrhea,[28-30] and in 1945, Keefer concluded: "Of all infections which respond to penicillin in a dramatic fashion, those due to gonococci head the list. Certainly 95 per cent of men and women with local gonococcal infections are cured within 24 to 48 hours after injection of 100,000 to 150,000 units in divided doses over a 12 to 15 hour period. When arthritis, peritonitis, or other systemic infections are present, 100,000 to 200,000 units daily for five to seven days or longer are required. Penicillin is the drug of choice in the treatment of all gonococcal infections."[29]

From 1942 through 1965, with the sulfonamides and penicillin both available for treatment, gonorrhea decreased at a rate of about 10 percent annually.[31] Unfortunately, from 1954 on, larger amounts of penicillin were necessary for growth inhibition of gonococci and for treatment of patients with gonorrhea than in previous years. The incidence of the disease began to increase in 1965, and evidence accumulated thereafter to show that at times 600,000 to 2.4 million units of penicillin were insufficient to cure the infection. The first penicillinase-producing strains of *Neisseria gonorrhoeae* were isolated in 1976, and by 1982 they had become endemic in many parts of the world. In the United States, the number of reported cases due to penicillinase-producing strains had risen more than tenfold between 1979 and 1981.[32-34] As the prevalence of penicillinase-producing strains increased, so too the dosage required to cure gonorrhea has increased.

Of importance throughout was the fact that 5 to 10 percent of men who develop urethral infection remain asymptomatic, do not seek medical help, and thus become carriers of the infection. In females, moreover, the disease is often mild and completely asymptomatic, the organisms may be carried for many months, and again the resultant reservoirs of infection keep case rates high. The ability of gonococci to persist, particularly in females, without producing symptoms thwarted even penicillin's capabilities. Despite the latter results, however, the striking ability of penicillin to inhibit growth of gonococci and to so rapidly cure gonorrhea in the early 1940s provided strong support for the belief that penicillin must be produced in quantity and made available for general use.

Thus, in the three years between 1942 and 1945, Chester Keefer, working under the auspices of the OSRD with clinicians throughout the United States, obtained abundant information on the beneficial effects of penicillin against staphylococcal, pneumococcal, streptococcal, gonococcal, and certain other bacterial infections.[35] But more was to follow.

The Treatment of Syphilis

A dramatic era in the chemotherapy of syphilis began at the Venereal Disease Research Laboratory [at the U.S. Marine Hospital on Staten Island, New York] in March 1943, when five rabbits with early scrotal chancres were given penicillin intravenously every four hours for four days. The disappearance of the *Treponema pallidum* from the lesions and the rapid healing of the ulcers were challenged with disbelief because the organisms had not been killed earlier in *in vitro* experiments.... Proof of the cure was confirmed later when the lymph nodes of the penicillin-treated animals were proved to be non-infectious. These findings warranted the early trial of penicillin in the treatment of early syphilis in man without further delay.[36]

On December 23, 1942, Dr. Harry Eagle, then affiliated with the Johns Hopkins University School of Medicine in Baltimore requested from the Committee on Medical Research a small amount of penicillin so that he might conduct a few in vitro tests of its activity against spirochetes.[37] Chester Keefer allocated 25,000 units for the tests, and on March 22 Eagle reported that penicillin had no effect on spirochetes in vitro. Eagle's findings were soon confirmed by Dr. John F. Mahoney at the U.S. Marine Hospital and Venereal Disease Research Laboratory on Staten Island. John Mahoney had tested the activity of penicillin against *Treponema pallidum* during the course of a study of the effect of penicillin on sulfonamide-resistant gonorrhea.[38] Later it was established that neither Eagle nor Mahoney had continued his experiment long enough to show the effectiveness of penicillin—a drug that requires active cell multiplication—against an organism that at best multiplies slowly.

John Mahoney's interest in syphilis had been aroused many years earlier, while on a U.S. Public Health Service tour of duty in Europe.[38] When he became director of the U.S. Marine Hospital in 1929, he began to devote his spare time to research on syphilis, and by 1943, he was thoroughly familiar with the course of experimental syphilis in animals and with the available methods for its treatment. He was by then familiar also with the toxicity and limitations of arsenotherapy, the main treatment then used for syphilis in humans.

Syphilis is caused by a spirochete, a corkscrew-like organism about six to twenty microns in length and known as *Treponema pallidum*. The disease may be acquired or congenital. If acquired, the spirochetes (Treponemes) enter the body through the skin or mucous membrane by bodily contact, usually through sexual intercourse. If congenital, the

152

organisms are transmitted to the fetus of the infected mother when the spirochete penetrates the placental barrier. From onset, syphilis is a generalized disease—the organisms invade all parts of the body. There are three stages of the disease. The primary stage lasts for varying periods of time, up to approximately seven to eight weeks after infection. During this period the primary lesion (chancre) appears at the initial point of contact. Gradually, the lesion heals even without treatment. The secondary stage begins while the primary lesion is healing, and within a few weeks or months, a rash appears, which may be accompanied by fever and some nonspecific symptoms. Open sores develop and contain large numbers of spirochetes which distribute themselves rapidly throughout the body. A period of latency may then follow, during which the organisms continue to penetrate all parts of the body. Finally, gummas, or granulomatous lesions of various organs, may appear, indicating the tertiary or late stage of the disease. Syphilis is chronic, with long periods of latency and a tendency to relapse.

The disease was highly prevalent in the United States in the early

VENEREAL DISEASES—Reported Cases of Gonorrhea and Syphilis (All Stages) per 100,000 Population by Year, United States,* 1919–1976**

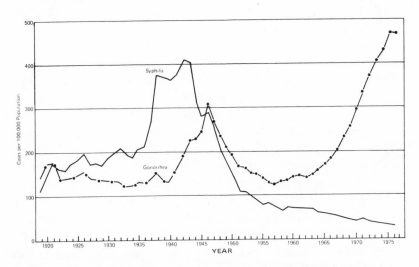

FIGURE 8.1. Venereal diseases—reported cases of gonorrhea and syphilis (all stages) per 100,000 population by year, United States, 1919–76. (From: MMWR Center for Disease Control: Annual Summary 1976 *[August 1977, vol. 25, no. 53], p. 67.)*

153

1940s and was of great concern to the armed forces (see figure 8.1). A total of 485,560 cases of syphilis (all stages), exclusive of known military cases, were reported to the U.S. Public Health Service in 1940, the case rate being 368.2 per 100,000 population in that year. In New York City alone, where fairly accurate records were kept from the time of Dr. William Halleck Park onward, 30,178 cases were reported in that year. Statistics provided by the Selective Service Commission on more than 15 million registrants for military service, moreover, indicated that approximately 5 percent of men of military age during World War II had some form of venereal disease at the time of physical examination for entry into military service.

The U.S. Public Health Service was especially interested in syphilis also, for their statistician for the Venereal Disease Service, Johannes Stewart, had recently conceived a program for curing an alleged 1 million cases of syphilis in Trinidad. He had proposed to the Office of Strategic Services that the U.S. Public Health Service set up a plant for manufacture of penicillin in Trinidad, the penicillin to be used entirely for the treatment of syphilis, and had proposed also that the Cancer Institute of the U.S. National Institute of Health at Bethesda be converted into a penicillin factory. Whatever may have been the merit of Stewart's proposals, the evaluation of penicillin in the treatment of syphilis was undertaken, although on a very different basis.[37-42]

In July 1942, the U.S. Army, acting on the recommendation of the Subcommittee on Venereal Disease of the National Research Council, adopted for routine treatment of early and latent syphilis a schedule that involved administration of forty injections of bismuth subsalicylate in a period of twenty-six weeks. This treatment was used in the early 1940s and, impractical though it was, there was no alternative—not until penicillin was introduced.

During the course of a study of sulfonamide-resistant gonorrhea, Drs. John Mahoney, Richard C. Arnold, and A. Harris in June 1943 first tested the effect of penicillin on the course of lesions of rabbit syphilis. They obtained their penicillin by growing the mold in their own laboratory and produced enough to clearly establish its spirocheticidal activity in animals. The spirochetes disappeared rapidly from lesions, and this limited experience convinced them that treatment of a few patients was warranted.[36,38,39]

In the same month, with penicillin obtained through war procurement agencies, "four male patients with darkfield positive primary syphilis received intramuscular injections of 25,000 units of amorphous penicillin at

PLATE 8.4. *"Penicillin Pioneers—Modern Magic Bullet Discovered in Penicillin." (Reproduced from: the* New York World-Telegram, *May 1, 1946.)*

four-hour intervals, night and day, for eight days, or a total of 1,200,000 units [each]. In October 1943, after a series of careful darkfield, serologic, and clinical examinations on these patients, Dr. Mahoney reported before the annual session of the American Public Health Association that penicillin apparently possessed powerful antisyphilitic properties and, as an added advantage, promised nontoxicity combined with brevity of administration to an exceptional degree."[36]

I have a mental image of the room where I first heard Mahoney and his associates describe their results on the use of penicillin in the treatment of syphilis. The room was crowded. Loudspeakers and projection equipment were not as sophisticated then as now. Everyone strained to hear what was said, and the impact was electrifying. By then much had been

155

written on the activity of penicillin, but no one had expected that an antibacterial agent would be active against spirochetes as well. Hearing John Mahoney describe the effect of penicillin on the course of syphilitic lesions was overwhelming.

A preliminary report on Mahoney and Arnold's observations was published in 1943 in *Venereal Disease Information*,[43] and in the September 9, 1944, issue of the *Journal of the American Medical Association*, Mahoney described their findings in greater detail, presenting follow-up observations for periods in excess of 300 days as well as preliminary observations on an additional 100 patients.[44] In the same issue, Dr. J. Earle Moore and his associates at the Johns Hopkins University Medical School[45] and Dr. John H. Stokes in Philadelphia,[46] separately reported preliminary results in the treatment of 1,418 patients with early syphilis and 182 cases of late (mostly neurosyphilitic) disease. Army, navy, Public Health Service, and civilian hospitals had pooled their efforts to produce this amount of information in so short a time. It was the beginning of what was probably the largest and most effective cooperative study ever conducted.[47-49]

Dr. Richard C. Arnold, who was affiliated with John Mahoney from the beginning, later described this eventful period:

During a two-week period beginning June 21, 1943, four young men with darkfield positive, sero-reactive primary syphilis volunteered for an historical experiment. Each man was given intramuscularly 25 thousand units of water soluble sodium penicillin every four hours for eight days. The size of the individual penicillin dosage was dictated by the fact that the ampoules contained 25,000 units. The injection schedule was comparable to [that used successfully in] animal studies. At my insistence, continuous therapy was administered long enough (at least one week) to eliminate the infecting organism from the human body. It is important to recall that the curative effect of one week of intensive arsenical therapy for syphilis had already been demonstrated in extensive animal and human studies. However the fatality rate from drug toxicity had rendered that arsenical regimen impractical.

With the experimental and clinical knowledge of arsenical therapy as a background, we had an adequate body of knowledge with which to begin the crucial experiment with the new non-toxic penicillin in small groups of animals and men.

The following sequence of events occurred in the development of the new concept of anti-syphilitic treatment. The causative factors that produced variations in the therapeutic response of human syphilis to penicillin were sought. A detailed study was made of each patient during the first 48 hours of treatment in order to record the clinical manifestations of the disease and any immediate or delayed reactions to therapy. Darkfield studies, made every two hours, revealed the rapid diminution in the number and motility of the *Treponema pallidum*, with complete disappearance of all forms in less than twenty hours. Within the first four hours of treatment the primary lesion of each patient developed acute redness, edema and oozing of serum. This changed in 24 hours to a rapidly healing, clean ulcer, despite repeated tissue irritations necessary for the Darkfield studies. In one patient, a transient macular eruption appeared for less than six hours. In all patients the mild toxic manifestations of temperature elevation and malaise were present for less than 24 hours, and these were replaced by the exhilarative effects of a treatment that quickly removed the highly infectious organisms of syphilis from the body. Quantitative serological studies by a battery of approved serological tests demonstrated conclusively each week the return of the elevated reagin titer to the negative phase. A favorable therapeutic effect in man was clearly indicated by these serological and clinical responses.

The disease's long unsatisfactory history of multifarious treatment schedules made it mandatory that the new penicillin treatment be evaluated carefully. It was imperative that the preliminary results in animals and in man be studied objectively; that proof of cure be established on a sound scientific basis; and that a simplified yet effective regimen be established quickly. Basic factors of adequate total dosage, frequency of injections and duration of continuous treatment, all had to be given thoughtful consideration. However, we had to keep in mind also that certain modifications might improve the therapeutic results, and that too many variations would result in a chaotic state of changing therapy that could never be evaluated properly.

The first study of one hundred patients indicated that 1.2 million units divided into sixty equal doses, given every three hours, would cure many individuals with early syphilis. However, a few treatment failures indicated the need for increasing the dosage while maintaining the same time schedule. The second regimen of 2.4 million units produced more satisfactory clinical responses, and we were convinced that penicillin would cure early syphilis.

The comparability of response to drug therapy in rabbits and in man, plus the relatively rapid procedures for determining a bacteriological cure in rabbits, were the significant facets of the entire research process.[50]

In June 1944 penicillin was adopted as the drug of choice for the treatment of early syphilis in the U.S. Army's European theater of operations. A few months later, it was adopted as the drug of choice by the British Royal Navy, Army, and Air Force—all this within two years of the time when the U.S. Army, upon the recommendation of the Subcommittee on Venereal Diseases of the National Research Council had adopted for routine use a regimen employing bismuth subsalicylate. Despite the chronic nature of the disease and the need for long-term follow-up to establish efficacy of the drug, sufficient information had accumulated to change medical opinion. Since it was clear, however, that a small but significant number of relapses were to be expected following penicillin therapy, the recommended dosage for treatment of early syphilis was fixed at 2.4 million units per patient, to be administered within 7½ days (sixty injections of 40,000 units every three hours).

The importance of animal experimentation in establishing the efficacy of penicillin in the treatment of syphilis cannot be overestimated. What could be expected with each dosage regimen, the likelihood of relapse, the significance of reinfection and of drug resistance, and the need for long-term follow-up—all were established in experimental animals. This minimized the unknowns when treatment of human subjects was undertaken and speeded the acceptance of penicillin in lieu of arsenicals as the standard form of therapy. As R. C. Arnold has written:

There is a great story in how and why the first syphilitic rabbits were treated... how they were used in experimental studies to prove the therapeutic effect of each new or modified [dosage] schedule before it was ever used in humans.... It was not easy to convince the syphilologists that penicillin was an effective curative drug, that it worked rapidly, that patients could be reinfected a few days after treatment had been completed. No penicillin-resistant strain of *Treponema pallidum* has ever been demonstrated, and it [penicillin] has now become the best prophylactic drug ever described, although the substance does not kill the treponemes on direct contact.[50]

Within two years after penicillin had been declared the drug of choice for the treatment of syphilis, however, problems arose.[48,51] Physicians found it necessary to give larger and larger doses of penicillin to cure pneumonia, to alleviate osteomyelitis, to control gonorrhea and syphilis. Although penicillin production had increased and yields per milliliter of culture fluid were far higher than in 1944 and before, the drug nonetheless was not as effective. The development of high-yielding strains of penicillia had led to the formation of new forms of penicillin.

Dr. Milislav Demerec at the Carnegie Institution in Cold Spring Harbor, New York, had produced by X-ray mutation a strain, X-1612, which from 1944 on was widely used for commercial production. It enormously increased the amount of penicillin produced. Subsequently, Myron P. Backus and J. F. Stauffer, botanists at the University of Wisconsin, went further and subjected the strain produced by X-ray mutation to ultraviolet light.[52] From this came another penicillium strain, designated Q-176, that produced still greater quantities of penicillin. What was not realized at the time was that Q-176 was producing a form of penicillin with remarkably high activity against strains of *Staphylococcus aureus* but lower activity (than that shown by previously manufactured penicillin) against strains of pneumococci, streptococci, the treponemes, and certain other organisms.

Not until December 1945 did it become public knowledge that penicillin was produced in multiple chemical forms, depending on the mold strain used and the culture medium employed. At that time, it became known to biologists and clinicians that penicillin could be produced as penicillin F, penicillin G, penicillin K, and penicillin X. Wartime secrecy had previously precluded broad dissemination of this information, although it was well known to those concerned with studies on the chemical structure of penicillin. The strain Q-176 produced high concentrations of penicillin K which had remarkably high activity against staphylococci. Earlier strains had produced primarily penicillin G (benzyl penicillin).

First evidence of the altered activity of commercially manufactured penicillin was obtain by Dr. Alan M. Chesney, dean of the Johns Hopkins Medical School and a well-known investigator.[48] He was asked to test the so-called K form of penicillin against syphilis in rabbits, on the assumption that the changed behavior of the commercial penicillin might be due to changed proportions of the various forms of penicillin in the product. He injected a virulent strain of *Treponema pallidum* into rabbits, waited for syphilis to develop, and then injected penicillin K into the animals in a dose that he assumed would be sufficient to cure. But even ten times that dose was not sufficient. Mahoney and Arnold made similar observations and Harry Eagle noted that the concentration of penicillin in the blood of the treated animals was only a fraction of that which could be demonstrated when penicillin G was administered.

By this time, I had become affiliated with Chas. Pfizer & Co., and—unaware of the observations being made by the syphilologists throughout the country—we had simultaneously noted in my own laboratories at Pfizer that in experimentally inoculated mice "impurities" in commercial penicillin seemed to enhance its antibacterial activity.[53] Later it became clear to us that the increased (or decreased) activity of commercial penicillin preparations was due to the proportion of penicillin G, K, X, or F present. Further, it was established by Ralph Tompsett and Walsh McDermott[54] at the Cornell University Medical College in New York that the low activity of penicillin K demonstrable in serum following

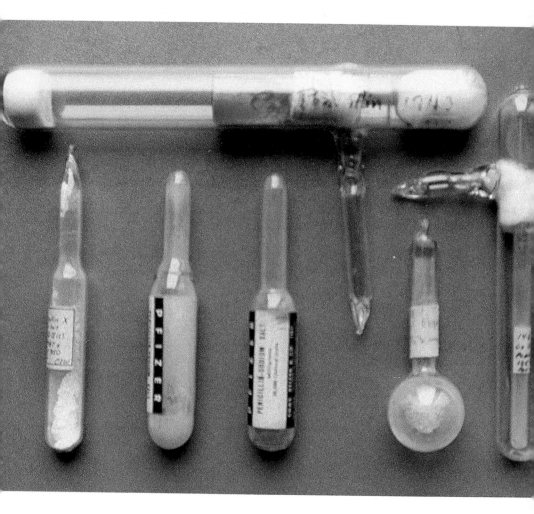

PLATE 8.5.   *Crude (impure) pigmented penicillin contrasted with modern-day crystalline penicillin G (benzyl penicillin): (a) upper left, first row, pure crystalline penicillin G manufactured in 1984 (Pfizer, Inc.) versus (b) fourth row, heavily pigmented crude penicillin prepared at Pfizer during early 1940s and (c) fifth row, pigmented calcium salt of penicillin prepared by Dr. Karl Meyer (Columbia University) in 1943; (d) upper right, first row, impure pigmented penicillin X (para-hydroxybenzyl penicillin), 170 units per milligram versus (e) third row, sodium salt of penicillin X, 950 units per milligram, prepared in May 1945, showing no pigmentation; (f) ethyl ester of impure penicillin prepared in 1942, showing yellow-brown pigmentation. Photographed with permission of Dr. Karl Meyer and Pfizer, Inc.*

administration to animals or humans (and its diminished efficacy) was due to the drug's capacity to bind to serum proteins.

Obviously, it was essential to minimize the amount of penicillin K produced by the penicillin-producing molds then in use. This was accomplished primarily by the use of precursors, chemicals that served as starters and encouraged the mold to produce more of the desired form of the drug and less of the unwanted forms.[55]

Penicillin X seemed more active against gonococci, the cause of gonorrhea in humans, and against *Streptococcus viridans*, organisms closely related to hemolytic streptococci and frequently responsible for cases of subacute bacterial endocarditis. For a time, penicillin X was produced by some manufacturers but its instability precluded its continued use. By incorporating derivatives of phenylacetic acid and other related chemicals into culture media, however, multiple other penicillins were produced.

Subacute Bacterial Endocarditis

The evaluation of penicillin in the treatment of acute bacterial infections stemmed largely from Howard Florey and Norman Heatley's visit to the United States in mid-1941 and from the observations made at Oxford prior to that time. In large part, the evaluation of the substance also stemmed from wartime pressure demanding that every possible means of combating infection in battle casualties be explored. Moreover, it was necessary to know if the quantity of penicillin Florey requested from the U.S. pharmaceutical industry could be justified to those whose invested monies would be used for its development and production.

The impetus for the evaluation of penicillin in the treatment of subacute bacterial endocarditis derived differently. It came from the humanitarian instincts of a few—perhaps ten or a dozen—persons who were simply concerned for those afflicted with the disease.

Ernst Chain once remarked that in actual practice he did not know of a major disease for which a cure had been found because the research worker wanted to find the cure. But Chain was a chemist who at the time may not have been well informed about drug and vaccine development. Henry Dawson, the first person to administer penicillin to patients, hoped to obtain a therapeutic response. He started his study of the substance specifically in an attempt to cure patients with subacute bacterial endocarditis.[56]

From 1940 on, I carried penicillin-producing cultures of a strain of *Penicillium notatum* to the Presbyterian Hospital wards each day so that patients with subacute bacterial endocarditis could observe with us the golden yellow surface droplets of penicillin-containing liquor exuding from the feltlike layer of mold in the culture flasks. The patients became interested. They saw the assay tubes and they saw the plates indicating the activity of the substance against their own infecting organisms. Their hopes rose high, even though they were told that the substance probably would never help them, for they were confident that it might help others who came after them.[56,57]

It took little time to convince us that a chronic infection such as subacute bacterial endocarditis would require high dosage and long-term administration. The amount of penicillin available was small, and almost two years elapsed before the first patient could be adequately treated. Even then, the response was slow for treatment was frequently interrupted. The drug was still in short supply. But, on October 17, 1942, a twenty-seven-year-old woman with classic signs of subacute bacterial endocarditis was admitted to the Presbyterian Hospital in New York City. After the administration of 1,420,000 units of penicillin over a period of more than two months, the bacteremia (presence of viable microorganisms—in this case streptococci—in the circulating blood) appeared to have been controlled.[20,56,57]

H.H., a woman aged 27, known to have had "endocarditis" at the age of 10 was admitted to Presbyterian Hospital on October 17, 1942, with a history of weakness, loss of weight, palpitation and low grade temperature of two months' duration. Three months previously some teeth had been extracted, and one month later the tips of several fingers became painful. Examination showed several petechiae [minute hemorrhagic spots] in the mouth, a moderately enlarged heart with aortic and mitral systolic and diastolic murmurs, and a palpable spleen. Three blood cultures were positive for [showed the presence of] *Streptococcus viridans* with 60 to 80 colonies per cubic centimeter (i.e., per cubic millimeter]...the organism ...*in vitro* was one-half as sensitive to penicillin as the standard [highly susceptible] strain of hemolytic streptococci [then commonly used for comparison].

Penicillin, 10,000 units every three hours intravenously, was started on October 23 and continued for five days. Blood cultures on November 10 and November 14 were positive [showed the presence again of bacteria in the blood]. Penicillin was again started on November 17 and continued for a period of six days. Following this third course blood cultures became negative [sterile] but, because of the persistence of a low grade fever, therapy was resumed as a precautionary measure. From December 7 to December 12, 5,000 units were administered every three hours intravenously. During this time the temperature rose slightly and petechiae appeared on the finger tips. In spite of these signs, blood cultures were sterile and continued to remain so. The total amount of penicillin administered was 1,420,000 units.

Although the bacteremia appeared to have been satisfactorily controlled, the patient developed a number of serious complications during the ensuing weeks [with resultant radical pelvic surgery in April 1944].... Throughout this period, however [from February 5, 1943 to April 6, 1944], nineteen blood cultures were taken and all were sterile. [When seen in January 1945, the patient was leading a normal active life.][56]

Dawson subsequently treated a total of five patients with subacute bacterial endocarditis during 1942 and 1943. The outcome in two was successful. In the remaining three, the response was satisfactory as long as, but only as long as, penicillin was administered. Two of the three later received more intensive therapy, with completely satisfactory results. The third died of a cerebral accident six weeks after discontinuance of penicillin

PLATE 8.6.   Dr. Thomas Hunter of the University of Virginia, who, with Dr. M. H. Dawson, treated some of the first subacute bacterial endocarditis patients with penicillin. Shown with Dr. Mercy Heatley at the 11th International Congress of Chemotherapy and the 19th Interscience Conference on Antimicrobial Agents and Chemotherapy, Boston, Massachusetts, 1979. (Courtesy of the American Society for Microbiology.)

therapy. Postmortem examination of the heart revealed only slight evidence of residual infection.

Subsequently Dawson, in collaboration with Dr. Thomas Hunter, later professor of medicine at the University of Virginia School of Medicine, treated another fifteen patients with subacute bacterial endocarditis with penicillin. Two of the fifteen succumbed, each apparently because of a cerebral embolus. All others became clinically and bacteriologically free of infection.[56]

Six of seven additional patients treated still later by Dawson and Hunter (but prior to Junary 1944) responded similarly. It was clear that a cure for subacute bacterial endocarditis had been found. Yet in August 1944, Wallace Herrell at the Mayo Clinic reported the administration of penicillin to four patients with the disease, two of whom were infected with nonhemolytic streptococci and two with a "micrococcus." All four were classified as treatment failures although the blood stream of one had been temporarily sterilized. Herrell concluded that the "usual doses" of penicillin were ineffective, and many others still believed this to be true.

A variety of microorganisms may cause subacute bacterial endocarditis, the most frequent being nonhemolytic, or green (viridans) streptococci. Of low virulence and usually present in the mouths of most humans, they may occasionally invade the bloodstream, following such trauma as, for example, a tooth extraction. Once in the blood, they remain harmless unless the heart valves are damaged or scarred, as may be the case in patients who have had one or more attacks of rheumatic fever. In such patients, the infecting organism, carried in the circulating blood, may establish itself on a heart valve where it is well protected from the action of phagocytic cells and where conditions are generally optimal for its growth. There it grows, multiplies freely, and colonizes. The disease progresses insidiously, pursuing a long clinical course lasting several months. Its distinguishing feature is its chronicity and its high mortality when untreated. Onset of the disease is accompanied by low-grade fever and malaise; petechiae may eventually appear in the skin and mucous membranes; and there may be pain in the muscles and joints.

The microbe guards itself cleverly from medical attack by burrowing into...the heart valves and covering itself with a cauliflowerlike vegetation of clotted blood. In this...nest it grows and swarms, seemingly out of reach of any curative serum or chemical. Then it sallies out into the blood. It not only wrecks the heart by attacking the delicate valves but it causes...mischief all over the human body. Bits of blood clot from the heart valves detach themselves, swirl through the circulating blood, and lodge in the arteries of the brain, the kidneys, the eyes, the skin, the lungs and the heart itself. This blocking of arteries, called embolism, devitalizes one part of the body after another.[58]

By August 1943, Chester Keefer, the physician responsible to the War Production Board for distribution of available supplies of penicillin, had laid down guidelines for use of the scarce and valuable substance. Subacute bacterial endocarditis, which they knew would require high dosage and long-term administration, was designated unsuitable for treatment with the drug. But this was a serious infection, and not all agreed to abide by Keefer's edict. As small amounts of penicillin gradually became available, some was used to treat patients with subacute bacterial endocarditis.

Although no penicillin was officially released for the treatment of subacute bacterial endocarditis, Karl Meyer and I continued to make some of the drug in our own laboratories until early 1944, largely for Dawson's clinical use. Moreover, through an affiliation with Chas. Pfizer & Co., additional supplies became available at an early date. Some patients were treated with low dosages. The drug was administered only intermittently, for supplies ran out often. This may have been fortuitous, however, for this intermittency may well have allowed the infecting organisms the opportunity to multiply, thus regaining their full susceptibility. Whatever the reasons, some patients responded well.[57,59,60]

In early 1943, a group of doctors at the Jewish Hospital in Brooklyn, headed by Dr. Leo Loewe, began to acquire further evidence supporting the belief that penicillin could cure patients with subacute bacterial endocarditis. For some time, Loewe had been studying methods for the production of experimental subacute bacterial endocarditis in rabbits and the effect of sulfonamides, administered in combination with heparin (an anticoagulant), on the course of the disease. He was convinced that heparin was essential to successful treatment in both animals and man. But he had difficulty maintaining adequate concentrations of heparin in the circulating blood of humans. Through Dr. Ralph Shaner of Roche-Organon in Nutley, New Jersey, Loewe learned of a carrier known as Pitkin's Menstruum, containing gelatin, glacial acetic acid, and glucose, which he believed would slow the elimination of water-soluble compounds from the blood. Heparin was water soluble. Loewe tried this menstruum, first in rabbits and with the sulfonamides. He became convinced of its merit. Later he maintained that penicillin could cure patients with subacute

165

rk World-Telegram 5/2/46 21   SECO

PENICILLIN PIONEERS

Dreaded Heart Infection Yields to Teamwork

Fourth of a series.

By FREDERICK WOLTMAN,
World-Telegram Staff Writer.

To a Brooklyn physician and a Brooklyn chemical manufacturer who owns a fourth of the Brooklyn Dodgers goes the credit for one of the most dramatic chapters in the history of medicine's new wonder drug—penicillin.

These two men proved to the medical world that penicillin cures that dreaded disease of the heart valves—subacute bacterial endocarditis—and they revolutionized the treatment of an infection which had been virtually 100 per cent fatal.

Their discovery was made against overwhelming odds and in the face of published reports that the new drug had failed as a cure. It was a triumph of teamwork between science and industry.

The principals in this little-known drama of science are Dr. Leo Loewe and John L. Smith, president of Chas. Pfizer & Co., Inc., a Brooklyn chemical firm. The results of their work have been published in numerous medical journals.

For decades, physicians the world over had been trying to find a cure for subacute bacterial endocarditis, a disease in which germs attack the heart valves. In the early course of the infection, the victim is not very sick. He feels fatigued, loses a little weight and runs a low fever. Within a few months—a year at the most—he develops paralysis, apoplexy, blindness, eventually heart failure.

The disease was a horrible one, with death inevitable, and up to that time medicine was helpless in coping with it.

The sulfa drugs then came along and saved a few lives, but less than 2 per cent, according to New York Hospital, which reported three cures out of 161 cases.

Produced in Rabbits.

Meanwhile, for the last 25 years, Dr. Loewe had been especially interested in the ailment. He produced it artificially in rabbits for experimental purposes. Associated with him in the investigation at Brooklyn Jewish Hospital were Drs. Max Lederer, Philip Rosenblatt, Erna Alture-Werber, Harry [...] and Matthew D. Levin

death who had already been given the last rites of her church.

Still other sufferers, treatment failures elsewhere, were brought to Brooklyn Jewish Hospital from hospitals all over the United States and Canada, some of them in so critical a state that oxygen was administered during the trip to keep them alive.

Dr. Loewe was achieving what J. D. Ratcliff described in "Yellow Magic. The Story of Penicillin," as "the most glowing results ever recorded in this fearsome disease."

The cures mounted to the scores in what was becoming a miracle

Above, Laurette DesRosier, a subacute bacterial endocarditis patient at Brooklyn Jewish Hospital, partakes of a meal while penicillin is administered by slow drip technique. Below, Al Tutunjian sits up and reads while undergoing treatment.

John L. Smith (above), president of Chas. Pfizer & Co., Inc., of Brooklyn, a principal in an epochal drama of penicillin.

World-Telegram Photos by Hingson and Orens.

PLATE 8.7. *"Penicillin Pioneers—Dreaded Heart Infection Yields to Teamwork." (Reproduced from: the* New York World-Telegram, *May 2, 1946.)*

bacterial endocarditis only if it were administered in combination with heparin in Pitkin's Menstruum. Although Loewe's own data eventually showed that neither heparin nor Pitkin's Menstruum was essential for penicillin treatment of the disease, his early observations with this combination of drugs were impressive.

Loewe, in June 1943, was asked to treat the daughter of another Brooklyn doctor, a friend who had heard of his interest in subacute bacterial endocarditis.[61] It was obvious that the child was suffering from subacute bacterial endocarditis. Initially Loewe gave her large doses of sulfadiazine, along with heparin in Pitkin's Menstruum, but the results were unsuccessful. He needed penicillin, and he turned for supplies to a man well known in Brooklyn, John L. Smith—vice-president of Chas. Pfizer & Co. and part-owner of the Brooklyn Dodgers. Smith at the time was better known for his interest in baseball than in penicillin, an interest that created a strong bond between himself and Alexander Fleming.

Smith's interest in penicillin had been aroused previously by Henry Dawson, and he was well aware of Dawson's belief that penicillin would cure patients with subacute bacterial endocarditis if the drug could be administered in sufficiently large amounts and for a long enough period of time. He was aware too that Dawson had some evidence to support his belief. Since the summer of 1941, Pfizer had been researching methods of producing penicillin by surface and submerged fermentation and methods for its recovery and purification. It was important for Smith to learn if the substance could be produced in quantity and at low cost, without harm to other fermentation processes being carried out in the same vicinity. Pfizer was then involved in a variety of other fermentation processes, and it was a small firm with fewer than eight hundred employees, many of them with life savings invested in the organization.

When Dr. Loewe asked John Smith for penicillin for his friend's daughter, Smith knew that its use for treatment of a subacute bacterial endocarditis patient would be frowned upon by government authorities. Chester Keefer had just been appointed, on May 9, 1943, "guardian" of the small supplies of penicillin and his committee had issued guidelines for use of the drug. Smith had agreed, as had others, to turn over to the Office of Scientific Research and Development for clinical trials all the penicillin the firm produced—except a small amount that could be held for corporate research. John Smith went to see the doctor's young daughter who was suffering from subacute bacterial endocarditis. He could not resist. On June 19, 1943, he released penicillin to Loewe for treatment of the child, and on July 23, she was discharged from the hospital, apparently cured. Her infection had been due to a strain of pneumococci, highly susceptible to penicillin. The outcome of treatment was striking; it was reported widely in the lay press and had a profound effect on the future of penicillin production.

Loewe had also been asked in May of the same year to treat another patient with subacute bacterial endocarditis who had failed to respond to large doses of sulfadiazine administered with heparin in Pitkin's Menstruum. In this instance, the infecting organism was a strain of *Streptococcus viridans*, less susceptible to penicillin than pneumococci. Initial results with penicillin were poor, and Loewe decided that massive doses of the drug must be given if a satisfactory response were to be obtained. Again he turned to John L. Smith. On August 26, he started a program to evaluate the effects of massive doses of penicillin administered intravenously, usually by continuous intravenous drip.

Distribution of penicillin for use in the treatment of subacute bacterial endocarditis became virtually illegal after the July 16, 1943, WPB allocation order. The agreement with the OSRD and the WPB, however, was that each month Pfizer could retain 8 million units for the company's research program. Smith and his staff thereafter supplied the OSRD and WPB all the penicillin they asked for and more, and then used their 8 million units to help Leo Loewe's project.

Loewe subsequently treated a large number of subacute bacterial endocarditis patients with penicillin. His first seven cases were the most important, however, for they proved what the drug could do.[62,63] Admittedly some relapsed and they relapsed just at the time when the Committee on Medical Research was trying hardest to prevent the use of penicillin for these patients. Moreover, Shaner simultaneously terminated his agreement to prepare Pitkin's Menstruum for Loewe's patients. But Elmer Bobst, president of Roche & Co. and a friend of John Smith's, became interested; he again made the heparin in Pitkin's Menstruum available, and John Smith took a chance and supplied Loewe with the penicillin he needed.

Loewe confirmed all of Dawson's findings with respect to the effectiveness of penicillin therapy in the treatment of subacute bacterial endocarditis. He established, moreover, the value of administering the drug in high dosage and over prolonged periods of time, using intravenous drip when necessary to achieve the highest possible concentration of the drug in the body. He also established the nontoxicity of these high dosages. All this was confirmed by Dr. Ward MacNeal at the New York Post-Graduate Medical School and Hospital of Columbia University, where John Smith, who was interested in these patients, devoted many weekends to visiting with them.[64]

In late 1943, Chester Keefer visited Leo Loewe in Brooklyn to evaluate for himself the results obtained in the first seven subacute bacterial endocarditis patients that Loewe had treated with penicillin. In September 1945, he outlined in the *Medical Clinics of North America* what he called important information regarding nonhemolytic streptococcal subacute bacterial endocarditis. The infecting organism, he wrote, must be susceptible to penicillin in vitro; a daily dosage of at least 200,000 units should be used, with 300,000 units providing a greater margin of safety and even larger doses rarely necessary; and the duration of treatment should be at least two weeks and preferably three. In 50 percent of cases, the infection would respond promptly and apparently permanently to one course of treatment.[65] Thus, Keefer had admitted the value of penicillin in the treatment of subacute bacterial endocarditis, and his conclusion was supported by later findings of the Penicillin Clinical Trials Committee of the British Medical Research Council.[66] The committee, however, recommended in 1946 that at least 2 million units of penicillin be given daily for a period of at least six weeks.

Some years later, Chester Keefer commented that he was glad people had persisted in the use of penicillin for the treatment of subacute bacterial endocarditis; it was good to know that some patients with this disease had been given the opportunity to live.

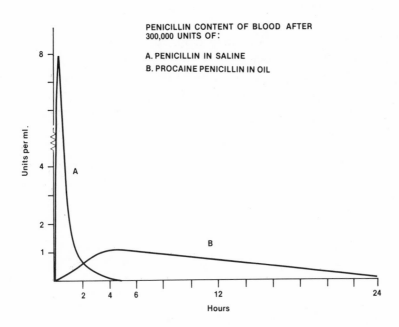

FIGURE 8.2.   Penicillin content of blood after administration of 300,000 units of penicillin (a) in saline and (b) in oil (procaine penicillin). (From: L. P. Garrod, Brit. Med. J. 2 [1950]: 453.)

The studies on penicillin in the treatment of subacute bacterial endocarditis and syphilis were initiated in 1940 and 1943, respectively, but not until 1944 to 1946 were the true merits of the drug in the treatment of these infections apparent. Only during these later years was sufficient penicillin available to permit the large doses and the long-term administration necessary to prove conclusively what it could achieve. By the end of 1943, however, it was clear that truly large-scale production was justified. No longer was there any question concerning penicillin's chemotherapeutic potential. The one remaining doubt stemmed from a persistent fear that chemical synthesis of the compound might be accomplished so rapidly that large-scale fermentation equipment would become obsolete too soon.[67]

Overall, these studies of penicillin, particularly in syphilis, had opened the door to modern penicillin therapy. They established penicillin G (benzyl penicillin) as the most stable and usable form of the drug. They called attention to the interrelationships between the chemical nature of penicillin and its biological properties, and they led directly to long-term studies on the pharmacological action and most useful and effective methods of drug administration.[68] More than ten years elapsed before the next really significant developments occurred. During these years, time-dose relationships, methods of drug administration, and long-acting forms of penicillin (penicillin in oil and beeswax and procaine penicillin, among others) were studied (see figure 8.2 and note 68). Also during these years, the efficacy of penicillin in the treatment of syphilis of the central nervous system was evaluated in the most carefully designed, large-scale, and controlled cooperative clinical trial ever conducted.

The place that penicillin would occupy in the physician's armamentarium in future years was clear. All that remained to be done was to make it available.

9

The U.S. Pharmaceutical Industry Meets the Challenge

Until July 1943, virtually all penicillin used for clinical purposes within the United States or by the armed forces was made by five companies—Merck & Co., E. R. Squibb & Sons, Chas. Pfizer & Co., Winthrop Laboratories, and Abbott Laboratories. By May it had become evident that penicillin would be needed by the military services in quantities far in excess of anything that had been imagined earlier. By then, it had become clear that, even though these five companies were producing significant amounts of penicillin, much effort was needed to bring their output to a level adequate to meet both civilian demands and those of the armed forces.

At a meeting of representatives from the U.S. War Production Board's Drugs and Cosmetics Branch, the U.S. Army and Navy, and the Office of Scientific Research and Development (OSRD), it was agreed that the War Production Board (WPB) would assume full responsibility for the production of penicillin in the months ahead. In the meantime, the OSRD would continue both its clinical evaluation of the drug and the chemical studies it had initiated with eventual chemical synthesis as their aim.[1-4] Although the industry, for some time past, had pooled its output of penicillin to arrive at a clear indication of its therapeutic efficacy, no concerted effort had been made to force increased production. It was time to do so, and in June the WPB launched a full-scale attack on the problem, setting a goal for the industry of 200 billion units of penicillin per month—an enormous amount by 1943 standards. Bear in mind that less than two years had elapsed since Howard Florey had requested that the U.S. pharmaceutical industry prepare 10,000 liters of culture fluid and

171

extract the penicillin so that he might confirm its effectiveness in a few patients.[5] Based on yields then available, his request was for a mere 5 million units or less.

More than 175 companies were investigated by the WPB to determine their ability to produce the needed quantity of penicillin. Twenty-two were selected to receive top priority on construction materials and other supplies necessary to meet the goal.[6,7] In order of their participation in the program, these 22 were as follows:

June 1943	Merck & Co., Inc.
	Chas. Pfizer & Co., Inc. [now Pfizer, Inc.]
	E. R. Squibb & Sons
	Winthrop Chemical Co., Inc.
July	Abbott Laboratories
	Hoffmann-LaRoche, Inc.
	The Reichel Laboratories [now Wyeth, Inc.]
	The Upjohn Company
September	Lederle Laboratories, Inc.
	Parke, Davis and Company
October	Eli Lilly and Company
	The Schenley Laboratories, Inc.
After	
December 1	The Ben Venue Laboratories
	Cheplin Laboratories, Inc. [purchased by Bristol-Meyers & Co. in 1943]
	Commercial Solvents Corporation [now International Minerals and Chemicals Corporation]
	Cutter Laboratories
	Harrower Laboratories, Inc.
	Heyden Chemical Corporation
	L. F. Lambert Company
	The MacLean Laboratories
	McKesson and Robbins, Inc.
	Sharp & Dohme, Inc.

When these firms were selected, their prior experience with penicillin, knowledge of fermentation methods for the production of chemicals, and experience with biological products were taken into account. None had adequate expertise in all three areas, but each had expertise in at least one. Help was enlisted from the Facilities Branch of the Chemicals Bureau of the War Production Board, from the Armed Service Forces, and from the OSRD. All worked in unison to increase production. On July 16th, an allocation order (No. M-338) was issued by the War Production Board which brought under the control of Fred Stock, chief of its drugs and cosmetics branch, the disposition of all penicillin that became available.[1-4]

Approximately 400 million units were produced during the first five months of 1943. In contrast, 425 million units were produced in June alone, and a steady increase thereafter brought the figure for December of that year to more than 9 billion units. Only two years had elapsed since Coghill and Moyer of the Northern Regional Research Laboratory in Peoria had reported the value of using a lactose–corn steep liquor-containing medium for production of penicillin. Already the U.S. pharmaceutical industry was showing signs that mass production was within reach.[4]

Major contributions of the War Production Board were to increase the number of participating manufacturing firms, to make available supplies and equipment needed for expansion, to ease the restrictions imposed on the industry by wartime regulations, and to funnel all supplies of the drug to the most important uses. Technical "know-how" already existed. Pilot plants were in operation at least at Merck & Co., E. R. Squibb & Sons, Chas. Pfizer & Co., and Abbott Laboratories. Both surface and submerged growth of the organism were being used for small-scale and pilot-scale production. The stage had been set.

How It Came About

Many claimed an interest in penicillin production prior to 1941; many claim to have been deterred only by the small amount of active material produced in fermentation liquors, the instability of the substance, and its cost. Those who read Fleming's first report in 1929[8] or the later reports from Harold Raistrick's laboratory[9] surely must have given momentary thought to penicillin's potential. But fleeting thoughts are not important; it is what is done that counts. It is easy to assume in retrospect that events of the past were the result of purposeful planning. But to prove this assumption is often difficult and at times impossible. So it is with the history of penicillin. The record is clear only from 1940 on.

As I have recounted, Henry Dawson, Karl Meyer, and I at the College of Physicians and Surgeons, Columbia University, initiated experiments in September 1940 to confirm the observations of Fleming and the Oxford group of investigators on penicillin's action and to produce on a small scale penicillin for use in the treatment of streptococcal and pneumococcal infections.[10] Dr. Dickinson Richards—who in 1956 received the Nobel Prize for his studies of pulmonary circulation—was also affiliated with the College of Physicians and Surgeons. His research laboratory was situated on the same corridor as ours, as were the laboratories of Michael Heidelberger, Alvin Coburn, and Alphonse Dochez, all interested in respiratory disease, particularly streptococcal and pneumococcal infections and the organisms that cause these infections. All were in close contact with one another and with some members of the staffs at Merck & Co., E. R. Squibb & Sons, and Lederle Laboratories (the last mentioned company was a producer of specific streptococcal and pneumococcal antisera). Karl Meyer, moveover, was a consultant to Schering & Co. We all talked freely of what we were doing in our laboratories.

How much of our enthusiasm for penicillin may have been conveyed to others and how much others may have stimulated us cannot be estimated now. It is unlikely, however, that anyone who knew us failed to recognize the importance we placed on the production of penicillin in amounts adequate for clinical use. Surely some, moreover, "made stabs" at growing penicillia (to produce penicillin) during the period from September 1940 to July 1941. Merck surely was interested in penicillin at an early date, as is apparent in a letter from Robert Coghill (chief of the Fermentation Division of the NRRL) to A. N. Richards, dated December 27, 1941. Scientists at Merck were so prepared to act promptly when asked by the government in October 1941 to consider the production of penicillin, that earlier studies must have been in progress:

Merck has a very beautiful set-up for the continuous extraction of penicillin from 1,000 liters of culture fluid per week. They...have plans drawn [moreover] and are acquiring apparatus to process ten times this amount of material.... Their concentrates are running approximately the same as...[those] of Heatley and Florey—40 Oxford units per milligram.[11]

As the reader will recall, our first studies on the fermentation of penicillin started at Columbia University on September 23, 1940. The culture of *Penicillium notatum* used on that day had been received from Roger Reid of the Johns Hopkins Hospital. Reid had received the culture directly from Alexander Fleming several years earlier.

The first series of flasks showed activity after four days and first sensitivity tests, using green streptococci [*Streptococcus viridans*] from cases of subacute bacterial endocarditis, were run on September 30, 1940. First toxicity and protection tests [to establish the drug's ability to protect against infection with streptococci] were run on October 2, 1940, using active broth filtrates. First concentrates were prepared by Dr. Meyer on October 7, 1940. First injection in patients on January 11, 1941.[12]

On October 28, 1940, Dawson wrote to Ernst Chain to acknowledge receipt of Fleming's strain of penicillium:

Prior to the arrival of your subculture I was able to obtain a subculture of Fleming's strain from Dr. Roger Reid (who is now in Baltimore) and we are now at work trying to prepare potent extracts. Your report has occasioned the very greatest interest over here and with the assistance of Dr. Karl Meyer we have been able to get started on the problem. I hesitate asking for information before your work is published but if you are prepared to release any data on the preparation of your extract, we are naturally very anxious to know about it. May I ask whether you get a better yield from synthetic media than with plain broth and has your material been used in the treatment of humans as yet? My inquiry is prompted by my desire to treat, if possible, a patient with subacute bacterial endocarditis in whom I am more than ordinarily interested.[13]

Chain's culture "showed purplish red sporulation on receipt and never produced any penicillin in our hands."[12] At Columbia, we continued work with Reid's strain of *Penicillium notatum*, evaluating methods of increasing the yield of penicillin in fermentation liquors and studying the nature of the antibacterial activity. Meyer meanwhile attempted its isolation and purification.

While this was transpiring at Columbia University, three commercial firms—Merck & Co., E. R. Squibb & Sons, and the Lederle Laboratories—were showing increased interest. The proximity of the New Jersey Agricultural Experiment Station to the Merck & Co. laboratories in Rahway, New Jersey, led to close contact in the 1930s and early 1940s between Dr. Selman Waksman and scientists at the Merck plant. Waksman had long been investigating microbial antagonisms in nature and had isolated a number of antimicrobial substances from cultures of microorganisms found in soil and elsewhere in nature; he had developed techniques for demonstrating their presence in microbial culture fluids. Waksman had become a consultant to Merck & Co. in 1938 and had sown the seeds for the company's subsequent interest in antimicrobial substances—an interest that spread easily to nearby E. R. Squibb & Sons.[14,15]

In May 1941 we reported our findings to the Society for Clinical Investigation.[16] From May 5 on, events occurred rapidly and almost

175

simultaneously. No longer was it possible to relate one to another sequentially, and no single event, in my judgment, triggered the surge of interest in penicillin that became apparent early in the summer of 1941. One might credit the lay publicity given to penicillin by science writers William Laurence and Steven Spencer in May of that year, the reaction of physicians to the medical report presented by Dawson to the Society for Clinical Investigation, or perhaps advance information on the content of the second Oxford report on penicillin which was published (in *Lancet*) in August 1941. Unquestionably, however, Florey and Heatley's visit to the States and the imminence of our entry into World War II were prime factors. Whatever the reason, there was an enormous ground swell of interest.

On June 6, 1941, G. O. Cragwall of Chas. Pfizer & Co. wrote to Henry Dawson:

We have been quite interested in the account of the paper on penicillin read by you before the America Society for Clinical Investigation. We noted particularly the report that your investigation has been hampered by difficulty in obtaining large quantities of solution containing the germicide.

It has occurred to us that possibly we might be of some assistance in this connection. As you may know, we probably produce more mold than any other firm in this country, our citric acid being produced by mold fermentation. In addition to citric acid we are also producing gluconic and fumaric acids by fermentative processes. We maintain a laboratory devoted to the study of such organisms and products of their growth, and are at the present time erecting a new laboratory building which will greatly enlarge our facilities. We are equipped to produce molds in large quantities and maintain pure cultures of a wide variety of organisms. Among those with which we have worked we might mention a large number of strains of penicillia, aspergilli, mucor, rhyzopi, and others. At the present time we are carrying out investigations designed to find other substances in our media in addition to the acids which we are producing commercially.

We feel that we are in an excellent position to produce large growths of any organism in which you are interested for use in your investigation. Should you be interested and feel that our assistance might be of some value to you, the writer will be quite pleased to discuss the matter further with you at your convenience.[17]

Initially little came of Pfizer's expression of interest. But one month later, Howard Florey and Norman Heatley arrived for their United States visit and were directed by Dr. Charles Thom to the new Northern Regional Research Laboratory (NRRL) in Peoria. Dr. Robert Coghill (who had spent thirteen years on the faculty of the Chemistry Department at Yale University) had just completed organizing and equipping the Fermentation Research Division at the NRRL.[18] Dr. Andrew Moyer, one of Coghill's associates, was assigned to work with Coghill on the penicillin project. Almost immediately, Moyer established that the addition of corn steep liquor to the culture medium increased the yield of penicillin tenfold. Moreover, the addition of phenylacetic acid (a degradation product from penicillin) to growing penicillin-producing cultures also increased yields. A strain of *Penicillium chrysogenum* isolated from a moldy cantaloupe produced unexpectedly high yields of penicillin under submerged growth conditions. Thus it became clear that changes in the composition of culture medium, strain selection, and even the use of precursors would alter the amount of penicillin produced. Finally, demonstration of the mold's ability to grow under submerged conditions in large vats indicated that at least some production problems were not insurmountable.[19]

All this was accomplished at the NRRL within four to five months after Florey and Heatley first arrived in July to enlist help. It was not accomplished, however, without previous experience.

Methods for the production of gluconic acid by strains of penicillia and aspergilli had been devised in the late 1930s by Percy Wells (see chapter 5) and members of the staff of Arlington Farms, a part of the U.S. Department of Agriculture then located where the Pentagon now stands in Arlington, Virginia. The pilot plant work was carried out by Arlington Farms personnel at the Agricultural Experiment Station at Iowa State College in Ames. A large aluminum rotary fermentor was employed. (Later, similar fermentors were used by a small firm in Kansas City for commercial production of calcium gluconate.) The project provided Wells with material for his doctorate dissertation at Georgetown University, and when Coghill moved from Yale College to Peoria in 1939 to organize the Fermentation Division of the NRRL, he inherited from Arlington the staff that had worked on the gluconic acid project.[19] Corn steep liquor had been used with success and they had learned techniques at Arlington for aseptic introduction of air into vats of fermentation liquor containing organisms requiring aerobic growth conditions. Thus, well before studies on penicillin production were initiated at the NRRL, knowledge had been acquired that later was essential to the penicillin production project. It was this experience that made possible NRRL's almost immediate success in the production of penicillin by fermentation.[18,19]

At the NRRL, gluconic acid was produced under research conditions only. The NRRL had had no experience in the conversion from research conditions to commercial production, although they knew that it could not be accomplished by simple multiplication of the amount of culture medium or inoculum, for example, in parallel with the shift in size of vat used. Pfizer, on the other hand, had made the transition from research to production in the mid-1930s. Among other things, it had learned methods for controlling the supply of air to aerobic organisms (such as penicillia) when cultivated under submerged conditions in large vats or tanks.

Research studies were begun in 1929 at Pfizer under Dr. James Currie, Jasper Kane, Alexander Finlay and Howard Carter on a process for the production of gluconic acid for the manufacture of calcium gluconate. Both submerged fermentation and surface fermentation were studied. Contrary to the situation in the production of citric acid, the submerged fermentation process proved to be very successful for production of gluconic acid; the surface fermentation process was a failure.

The successful development of a submerged aerobic fermentation in a stirred tank introducing sterile air and controlling pH was a revolutionary step forward in fermentation processes. The techniques and expertise gained in this development led directly to the successful deep tank fermentation process for penicillin some twelve years later in 1942.[20-22]

Pfizer's confidence in 1941–42 stemmed in part from the successful operation of its sugar–citric acid process and in part from its successful deep tank fermentation and commercial production of gluconic acid.

The United States Government and Penicillin Research

Pfizer, Squibb, Merck, and Lederle were the first companies contacted by Dr. A. N. Richards when in October 1941 he sought to stimulate interest in penicillin production.[23] In a letter to Dr. G. H. A. Clowes of Eli Lilly & Co. in Indianapolis, dated December 22, 1941, Richards summarized what had transpired:

Dr. [Baird] Hastings has told me of your expression of interest in current activities connected with the production and study of Penicillin in this country. It is to inform you of what has been taking place as well as to invite your cooperation that I am writing this letter.

In October I invited the Research Directors of four companies, Merck, Squibb, Lederle and Pfizer, to discuss with Dr. Bush and myself, representing the OSRD and CMR, Dr. Thom of the U.S. Department of Agriculture, and Dr. Wm. M. Clark of the Division of Chemistry and Chemical Technology of the National Research Council, with the view toward a cooperative effort to produce enough of the substance for further clinical tests and for chemical investigation. Three of the four companies represented had been approached by Dr. Florey during his visit to this country, and the fourth (Pfizer) was invited because of the activity of that company in commercial fermentations.

Little was accomplished other than the acquaintance of that group with the urgent desire of the CCMR [Chemotherapy Committee of the National Research Council] that production be expedited.

Another conference at which only representatives of the companies were present was held a month later, again with little result. On the 17th of December another meeting was held at which not only research men but also heads of companies were present. It developed that during the two preceding months the Northern Regional Laboratory of the Department of Agriculture at Peoria had succeeded in intensifying the production of Penicillin in laboratory cultures to a figure some 12 to 20 times that which had been accomplished in England and during the summer in Peoria. It became apparent that we would not have to wait long before supplies of the substance adquate for beginning [clinical] tests would be available.[23]

Richards's goal in 1941 was only to find one or more firms that would produce enough penicillin to meet Florey's immediate need. Developments at Pearl Harbor changed his goal, however, and the U.S. pharmaceutical industry soon found itself deeply involved in what was probably the greatest national and international collaborative venture of all time.

Merck, Squibb, and Pfizer took the lead in research on methods for production of penicillin and, later, in the push toward production on a scale large enough to permit clinical evaluation. They approached the problem with great optimism, even self-confidence, as is reflected in Dr. Baird Hastings's notes of December 18, 1941, which he wrote following a meeting held December 17 and attended by representatives of Merck & Co., E. R. Squibb & Sons, Lederle, Chas. Pfizer & Co., the U.S. Department of Agriculture, the National Research Council, and the Committee on Medical Research of the OSRD:

It was the sense of the meeting that with the facilities for production now in sight at Merck and Squibb, there will probably be an adequate supply [of penicillin] for the clinical testing and chemical needs at the present time. Dr. Coghill [of the NRRL] reported that they had achieved [a] twenty-fold concentration of penicillin, and he distributed among those present a report of their present method.[24]

From this time on, Merck and Squibb concentrated on enlarging their pilot plants, assuming that the drug would soon be synthesized and fermentation plants would be unnecessary. Their pilot plants were produc-

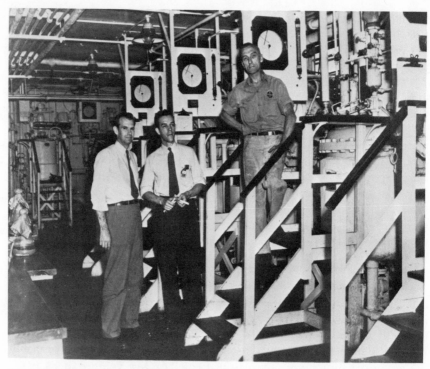

PLATE 9.1.   *Pilot plant fermentors for investigation of antibiotic production by submerged fermentation process at Merck & Co., Rahway, New Jersey, 1945. Left to right: Lloyd E. McDaniel, Bernard L. Wilker, unknown operator. (Courtesy of Merck & Co.)*

tive, and they led in supplying penicillin for clinical use. It was Merck's penicillin (12.5 gallons of fermentation liquor) that was shipped to Boston at the time of the Cocoanut Grove fire in December 1942.[25,26] Four hundred and ninety persons died in that fire; 174 others were injured or burned. Although it was not possible to evaluate just what penicillin accomplished, it was believed then—and is probably true—that the death toll would have been far greater had it not been for penicillin and its anti-infective power.

It was Merck's penicillin also that was used for treatment of Mrs. Ogden Miller, the first patient treated with penicillin supplied through the OSRD. Mrs. Miller's recovery did more, perhaps, than any other single event to justify efforts toward mass production of penicillin.[27-29]

From the beginning, Pfizer proceeded slowly and cautiously. It had

PLATE 9.2.   *Surface fermentation of penicillin in flasks at Chas. Pfizer & Co.,*
Brooklyn, New York, 1943. (Courtesy of Pfizer, Inc.)

few employees and few with scientific training; its sales had amounted to
only a little more than $7 million in 1940. Because it was a small firm that
had acquired expertise in fermentation technology, it had much to lose if
its knowledge were divulged too soon or if the penicillin-producing penicillium spores contaminated other products on which the company's viability
rested. Perhaps too Pfizer knew better than most others the pitfalls that
might be encountered in attempts to mass produce penicillin.[30–32]

No record has been found to indicate precisely when Pfizer first
undertook serious research on penicillin production. The first mention of
penicillin in the company's records appears in the summer of 1941.[33] In any
event, on October 2, 1941—six days before Dr. A. N. Richards's first
meeting with selected members of the U.S. pharmaceutical industry on
penicillin production—first samples of concentrated fermentation liquors,
some of them active against hemolytic streptococci in dilutions of 1 : 640 to
1:2,560 were sent from Pfizer to our laboratories at Columbia University.
From then on, fermentation liquors and samples for assay flowed to

181

PLATE 9.3.   *Oxford cup plate assay of penicillin devised by Dr. Norman Heatley at Oxford. Quality control of penicillin at Bioassay Laboratory at Chas. Pfizer & Co. about 1945. (Courtesy of Pfizer, Inc.)*

Columbia almost daily. By year end, Pfizer was supplying Columbia with 300 liters of fermentation liquor weekly, and we were urging the firm to find a way to concentrate or extract the penicillin from the crude liquor.[10,34,35]

The amount of penicillin produced by Merck, Squibb, Pfizer, and others in the industry increased steadily through 1941, 1942, and the first half of 1943 (see table 9.1).[3] By mid-1943, penicillin's value as a chemotherapeutic agent for control of infection was known to virtually everyone in England and the United States.[36] Now it was time to consider the project's original goal—that of production.

Many had realized from the beginning that production of penicillin would be economically feasible only if the producing organism could be grown in large tanks. Some had experimented with deep tank fermentation, even producing small amounts of penicillin. But all had continued growth of the organism in shallow layers in order to meet at least some of the needs of the medical profession and the armed forces.

Five commercial firms, working in close collaboration with one another, had been responsible for producing and making available virtually all the penicillin that had been used to establish its place in clinical medicine.[36] These firms—Merck, Squibb, Pfizer, Abbott, and Winthrop—had gambled the money, in some cases exceedingly large amounts, to produce enough of the drug to verify its value. The first three companies turned their entire penicillin production over to the Committee on Medical Research—at first for nothing, later at less than cost, and finally, when the amount produced became large, at commercial prices.[36]

A. N. Richards in 1943 still doubted that submerged fermentation would replace surface growth of the organism.[37] Nonetheless, on May 15, Jackson W. Foster and Floyd E. McDaniel of Merck & Co. filed a patent application (no. 2,448,790 which was issued on September 7, 1948) on a process for production of penicillin by aerobic submerged growth of strains of penicillia in media containing inorganic nutrient salts with molasses and/or corn steep liquor or brown sugar.[38] Coghill and Moyer had established earlier at the NRRL the advantages and feasibility of using submerged growth for penicillin production, but they did not file patent applications on their process until 1944 to 1945—although the results of their studies were known from 1941 on.[39-41]

Early in 1943, Pfizer's efforts to develop methods for commercial production of penicillin by submerged fermentation began to yield results, and in March, despite the nagging fear that other of the company's fermentation processes might be contaminated with penicillium spores, John Smith gave the signal to start increasing production. Research and production moved along hand in hand with J. H. Kane, G. M. Shull, E. M. Weber, A. C. Finlay, and E. J. Ratajak developing the fermentation process and R. Pasternak, W. J. Smith, V. Bogert, and P. Regna responsible for developing methods for its extraction. On May 10, Kane, then director of Biological Research and Development, reported progress but expressed the opinion that the submerged fermentation process was not yet ready for commercial use. On August 3, however, he requested that plans be made for construction and installation of a fermentor with a productive capacity of 1,000 gallons, and on August 27, Dr. E. J. Ratajak delivered penicillin fermentation liquor from a 2,000-gallon fermentor.[42-45] Three days earlier, the Works Committee had voted to lease outside space for manufacturing and to purchase equipment for

183

PLATE 9.4.   *Inoculating fermentation tanks with penicillin spores: inside the tanks is a rich nutrient broth, in which microorganisms feed and multiply rapidly (about 1943). When removed from the fermentors, the product is purified in a complex process and dried to a fine white powder. (Merck & Co., Rahway, New Jersey.)*

penicillin production to be installed in the new area. On September 20, the Rubel Ice Plant, located near the Pfizer plant in Brooklyn, was purchased with refrigeration equipment suitable for rapid conversion to a penicillin plant.[42]

Here, on March 1, 1944, the first commercial plant opened for large-scale production of penicillin by submerged fermentation. This was the plant that provided much of the penicillin used by the armed forces for D-Day casualties. Conversion of the plant had been accomplished under the leadership of John E. McKeen and Edward Goett in only five months. John Smith—then vice-president of Pfizer and a man who followed personally what went on in each experiment and each fermentation vat—had known by at least early August that supplies of penicillin would soon increase dramatically (see note 23). Howard Florey, when writing to John Fulton in August 1944 after visiting the Pfizer deep tank fermentation plant, described it as "a miracle of construction and organization."[46]

Pfizer had spent $950,000 in 1943 to expand the process for production of penicillin. Completion of the commercial plant for large-scale production and installation of necessary packaging and control facilities brought the figure to $2.98 million by the close of 1944. Pfizer bore these costs alone.[47,48]

It should be noted at this point that no single firm or individual was responsible for the successful development of penicillin. Many pharmaceutical firms, as well as the NRRL and some university laboratories, were involved in the penicillin research program. Reports from each were circulated freely to all others. As a Federal Trade Commission report summarized it, "It was specifically agreed that...the Chief of the Fermentation Division [of the NRRL] would advise the cooperating companies of progress made and [would] act as their consultant; and that, while the companies would proceed with the production research independently, this progress would be reported to the CMR [Committee on Medical Research] for such distribution as would advance the program."[49]

Robert Coghill, as chief of the Fermentation Division of the NRRL, collected and distributed the monthly reports that emanated from the participating laboratories and organizations. These are now on file at the U.S. National Archives in Washington, D.C., and provide a detailed account of all that transpired during the wartime period.[50]

The War Production Board Assumes Control

The War Production Board's allocation order of July 16, 1943, led to other firms being invited to assist in meeting production goals. At the time, seven companies—Merck & Co., E. R. Squibb & Sons, Chas. Pfizer & Co., Abbott Laboratories, Winthrop Chemical Co., Upjohn & Co., and the Reichel Laboratories (American Home Products)—were producing all the penicillin that was available. Six others were getting ready to produce penicillin and eight more were said to be "planning to get ready."[51]

Those involved in organizing a program to foster research on the synthesis of penicillin had added to this group the Wyeth Laboratories (a division of American Home Products), Parke Davis, American Cyanamid Co. (Lederle Laboratories), and Hoffmann-LaRoche. The War Production Board, interested more in the capacities of participating firms to produce than to carry out research, added the Ben Venue Laboratories, Cheplin Biological Laboratories (later affiliated with Bristol Myers Co.), Cutter Laboratories, Sterling Drug Company, and the Commercial Solvents Corporation. Not the least of these was the latter firm with its long experience in deep tank fermentation for production of industrial chemicals and the Chester County Mushroom Laboratories, which affiliated in late 1943 with Wyeth Laboratories.

Production soared in the last half of 1944. Pfizer alone, with the aid of its new submerged fermentation plant, increased its output from approximately 6.8 billion Oxford units in 1943 (a year in which both surface growth and submerged fermentation pilot plants were relied on) to more than 723.1 billion units in 1944. By June 1, 1944, Pfizer's production exceeded 70 billion units monthly. Nationwide production in March of that year (prior to the opening of the Pfizer plant) had been only 40 billion Oxford units. In June 1944, 117.5 billion units were produced, and in December 1944, production for the month was up to 293.4 billion units.[3]

Prices were slashed, making the drug available to more people (see figure 9.1). The WPB's allocation order, although placing the drug under strict control, opened the door to its use for treatment of civilian patients.[1,3]

Dr. Albert L. Elder, head chemical adviser, Chemicals Bureau of the WPB, was appointed coordinator of penicillin production in September 1943, with Dr. L. A. Monroe of the Office of Production Research and Development to assist him. It was a formidable undertaking to decide what equipment and supplies should be allocated to each company, what processes should be used, which firms were most likely to succeed. Fred

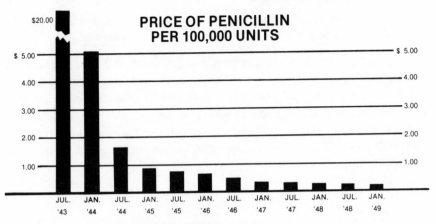

FIGURE 9.1.   *Price of penicillin per 100,000 units, July 1943 to January 1949.*
(From: Federal Trade Commission Report.)

Stock's and Albert Elder's success in making these decisions can be measured by the increase in available penicillin, from 425 million units in June 1943, just prior to the WPB allocation order, to approximately 40 billion units eight months later—all this before the first deep tank fermentation plant opened.[3,4,35]

In September 1943, the U.S. Food and Drug Administration for the first time requested every manufacturer to file a new drug application on penicillin and to start submitting to its agency a 25,000-unit sample of each batch of penicillin produced. Although it was acting without legal authority at the time, the Food and Drug Administration undertook to oversee the quality (biological safety, activity, and stability) of all penicillin produced.[52,53] In this informal manner, the Food and Drug Administration initiated a program that later became time consuming, costly, and unwieldy, but generally effective in controlling the safety and efficacy of all penicillin manufactured in the United States for clinical use.

The First Commercial Penicillin Plant

Pfizer's pilot plant in early 1943 produced up to 5 billion units of penicillin per month. The filtered fermentation broth in 500-gallon batches was adsorbed on activated carbon, eluted, and the eluate concentrated.

The chilled concentrated eluate was then acidified with sulfuric acid and extracted into chloroform. The solvent was dried over sodium sulfate and re-extracted with dilute sodium bicarbonate. This plant was in continuous operation to provide... material for clinical investigation and to fulfill Army requirements.[54]

Within less than six months after Jasper Kane had first recommended on August 3 that Pfizer make plans to convert to submerged fermentation, its pilot plant was supplanted by a commercial fermentation plant capable of producing billions of units of penicillin per month. This entailed the cooperation of all departments of the company and—more important—of the War Production Board, many other government agencies, and suppliers of equipment.

PLATE 9.5. The first submerged growth of mold occurs in tanks of several-thousand-gallon capacity. The tops of the tanks, which contain a corn steep liquor medium inoculated with penicillium, are shown here. After growth of the mold, centrifugal separators were used to separate the water-immiscible solvent from the aqueous phase containing the penicillin. (Chas. Pfizer & Co., Inc., Marcy Avenue plant, 1944.) (Courtesy of Pfizer, Inc.)

The firm's process had evolved from information acquired from others and from Pfizer's own experience both in the production of penicillin and in deep tank fermentation of other products.

The [deep tank] penicillin plant was 95% completed by the end of February and deep tank fermentation was initiated. Working 24 hours a day, seven days a week, the increase in penicillin production was dramatic.... 100 billion Oxford units of penicillin...[were] produced in July and 130 billion Oxford units...in August.... The...plant contained fourteen 7,000-gallon tanks...and sufficient...equipment [for] the recovery and purification of penicillin to a concentrated aqueous sodium penicillin solution.... With an average broth potency of 40 Oxford units per ml. and a 65% recovery, the plant was capable of producing 45 billion Oxford units of penicillin per month.... Throughout the plant some 250 employees were engaged in penicillin production.[55,56]

Modifications in the composition of the medium used increased broth potencies to an average of 100 Oxford units per milliliter of fermentation liquor, and changes in the recovery process resulted in production of penicillin with a potency of 900 to 1,000 units per milligram. By the summer of 1944, the price of penicillin had dropped from $20 in June 1943 to $2.20 per 100,000 Oxford units.

In early December, J. H. Kane advised John L. Smith that because of recent results, he believed that Pfizer could begin producing 300 billion Oxford units of penicillin per month in about ninety days' time. With a recovery of 75 percent, this meant a production of 225 billion Oxford units per month, five times more than the planned capacity of the new Pfizer plant. This prospect put the capacity of the drying and packaging equipment into focus as a limiting factor in penicillin production.

Pfizer ended the year having produced 723 billion Oxford units out of a total U.S. production of 1,600 billion Oxford units (see table 9.1).

What had made John Smith decide to gamble that Pfizer had at hand the necessary technology for mass production by submerged fermentation? Was he satisfied that Ratajak's success in a 2,000-gallon tank had been adequately confirmed? It is usually remarked that government pressure to increase production provided the impetus. But some of us who remember that summer and John Smith's intense interest in the patients being treated with penicillin in Brooklyn and Manhattan have always credited Patricia Malone for his decision to go ahead.[57] On August 11, this two-year-old child with staphylococcal sepsis was given "just seven more hours to live." One month later, she was fretting "with the restlessness of convalescence

189

PLATE 9.6.   *Sir Alexander and Lady Amalia Fleming with John E. McKeen, who was responsible for construction of the first plant for commercial production of penicillin by submerged growth of the mold. (Courtesy of Pfizer, Inc.)*

for the fast-approaching day when she [would] go home." The staphylococci that had invaded her bloodstream had been conquered by penicillin.[58] This child aroused the interest of many of us, but particularly of John L. Smith whose sixteen-year-old daughter had succumbed to an infection prior to the development of penicillin.

A Nationwide Transition to Submerged Fermentation of Penicillin

The WPB allocation order of July 16, 1943, and Pfizer's successful conversion to deep tank fermentation on a commercial scale in early 1944 settled the matter. Production now would involve submerged growth of the organism only. Merck and Squibb, who had done so much to make penicillin available for clinical research, were chagrined to find themselves temporarily left behind. They had gambled heavily on early synthesis of the drug and had assumed that there would be no need for commercial-scale fermentation plants.

Commercial Solvents Corporation, with its long experience in deep tank fermentation, took the lead. In like manner, E. R. Squibb & Sons moved quickly to erect a large fermentation plant for production of penicillin in tanks. Unfortunately Squibb's production was low throughout 1944 for contamination of their plant led to the destruction of penicillin—as fast as it was made—by penicillinase, an enzyme first described by Abraham and Chain at Oxford and manufactured by many strains of contaminating bacteria.[59-64]

The U.S. government built six production units at a cost of about $7.6 million. The companies operating two of the plants added nearly $2.9 million worth of packaging and power facilities. After the war, these plants were sold to the companies for a total of approximately $3.4 million. Sixteen other antibiotics plants were financed entirely by private industry at a cost of about $22.6 million. Of this amount, $14.5 million was approved for accelerated amortization under federal income tax regulations. This permitted the companies to deduct their investment from taxable earnings over a five-year period instead of the twelve- to fifteen-year period usually required for chemical manufacturing plants.[65]

Wyeth Laboratories and the Mushroom Industries Fill a Need

While Pfizer, Merck, and others were converting to mass production of penicillin by submerged growth of the organism, Wyeth Laboratories in West Chester, Pennsylvania, was asked by the WPB to utilize surface growth only for it was believed that this procedure would offer a sure supply of the drug.[66,67]

In September 1943 American Home Products Corp. merged seven wholly owned subsidiaries into a new Division known as Wyeth Laboratories which had become interested in penicillin production through Mr. G. Raymond Rettew, Technical Director of the Chester County Mushroom Laboratories. The Chester County Mushroom Laboratories produced pure cultures of mushroom spawn which was used as inoculum by growers of cultivated mushrooms. The company was the largest producer of such spawn in the United States and it had the equipment and the "know-how" to produce cultures of fungi in large quantities, precisely what was needed for production of penicillin. Large sterilizers, sterile inoculating rooms, air-conditioned incubator rooms, all were available. Its staff moreover had the necessary technical experience.

TABLE 9.1   Penicillin Highlights: The United States

	The United States Government Fosters Penicillin Research
July 1941	Studies on the production of penicillin began at the Northern Research Laboratory, USDA.
October 1941	The OSRD and Committee on Medical Research, National Research Council (NRC), called first meeting on penicillin. Merck, Squibb, Pfizer, Lederle represented.
December 1941	OSRD assumed responsibility, with cooperation of U.S. pharmaceutical industry, for evaluation of therapeutic efficacy of penicillin. NRC's Committee on Chemotherapeutic and Other Agents (Chairman: Perrin Long, M.D., January–June, 1942; Chester Keefer, M.D., June 1942–December 1945) responsible for allocation of the drug.
March 1942	First OSRD patient started on treatment with penicillin. From March 1942 until January 1943, all penicillin allocated by U.S. pharmaceutical industry to OSRD and Armed Forces.
January 1943	Penicillin purchased (at below cost to pharmaceutical industry) by OSRD through contract with Massachusetts Memorial Hospital (terminated December 31, 1945).
May 1944	OSRD research evaluation of penicillin's chemotherapeutic efficacy limited to special clinical situations only.
December 1945	OSRD evaluation of penicillin terminated. Results in treatment of 10,838 patients reported.

(Continues)

TABLE 9.1 Penicillin Highlights: The United States (*Continued*)

	1943: The U.S. War Production Board and the Pharmaceutical Industry
January to May	400 million units of penicillin produced in first five months of the year.
June	Pilot plants enlarged to produce 425 million units (by both surface and submerged fermentation techniques).
July	Allocation Order No. M-338 issued by U.S. government, bringing all penicillin in U.S. under War Production Board (WPB) control.
June to December	WPB increases number of firms that received wartime allocations of supplies and equipment for production of penicillin from five to twenty-two. Until July 1943, all penicillin used by OSRD had been made by Merck, Squibb, Pfizer, Abbott, and Winthrop Laboratories.
August and September	Pfizer moves ahead.
August 3	Initiates plans to construct and install fermentor of 1,000-gallon capacity.
August 27	Delivers penicillin fermentation liquor from 2,000-gallon fermentor.
September 20	Purchases Rubel Ice Plant for conversion to deep tank penicillin plant.
February 1944	Only seven months after issurance of the WPB Allocation Order No. M-338, the first plant for commercial production of penicillin by submerged fermentation technique opened by Pfizer with fourteen tanks of 7,000-gallon capacity.

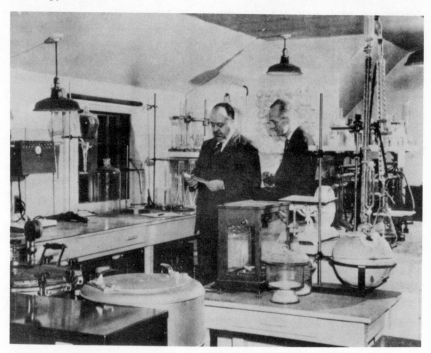

PLATE 9.7.   *At the Chester County (Pennsylvania) Mushroom Laboratories, later affiliated with Wyeth Laboratories, large amounts of penicillin were produced by surface growth of* Penicillium notatum.

Rettew recognized the similarity between what he had been doing for the mushroom industry and what was needed if penicillin was to be produced in quantity. Through Dr. E. B. Lambert, a research mycologist at the Bureau of Entomology and Plant Industry of the United States Department of Agriculture, Rettew contacted Dr. E. B. Fred, at the University of Wisconsin and then Dr. A. N. Richards, Chairman of the Committee on Medical Research of the Office of Scientific Research and Development. From the story Rettew told, Dr. Richards realized that he [Rettew] was right—there was a close relation between mushrooms and penicillia—and he accepted the Chester County Mushroom Laboratories as a full collaborator on the penicillin project.

Raymond Rettew obtained his first cultures of *Penicillium notatum* from Dr. Kenneth Raper at the Northern Regional Research Laboratories of the U.S. Department of Agriculture in Peoria, Illinois, and commenced work early in 1942. During the latter part of 1942, he worked alone, experimenting with methods for the cultivation of the Fleming strain and establishing the usefulness of 40-ounce culture flasks or short-necked bottles with four flat sides (flat, to facilitate stacking

194

the bottles in large numbers) then in use at the Chester County Mushroom Laboratories. In early 1943, new culture methods were introduced and the use of wetting agents and centrifugation solved the problem of emulsion which until then had plagued all penicillin producers.

No one at the Mushroom Laboratories was familiar with the requirements for the manufacture of drugs. Rettew therefore affiliated with Wyeth Laboratories, a well-established Philadelphia firm owned by American Home Products, a manufacturer of drugs, vaccines, foods and cosmetics. The penicillin-producing mold was grown in flat-sided, 40-ounce glass bottles, on shallow layers of culture medium containing corn-steep liquor and lactose as nutrients. Stacked on shelves in a sterile, air-conditioned room, they were inoculated with spores of *Penicillium notatum* and incubated for about seven days at 72°F. When the yield of penicillin was at its peak, the bottles were emptied into a tank and the mycelium (vegetative part of the fungus, consisting of interwoven fibers) was separated from the culture broth containing the penicillin. A total of 32,000 bottles were processed each day. Synthetic amyl acetate derived from fusel oil (a by-product of alcohol production in the beverage industry, called "banana oil" due to its odor and flavor of bananas) was used to extract the penicillin.

The penicillin solution was extracted from the solvent with sodium bicarbonate ($NaHCO_3$) on a buffer and the sodium salt of penicillin was then frozen in bottles and the water evaporated off under high vacuum to leave a residue of dry, yellow powder. In 1943, the purity was about 12 percent but the toxicity of both the penicillin and its impurities was low. The product was distributed under the label of the Reichel Laboratories (in Kimberton, Pennsylvania, a subsidiary of American Home Products and a producer of dried blood plasma). First deliveries were made to the U.S. Government in late June 1943.

Wyeth used only the cumbersome surface culture method of producing penicillin, knowing that ultimately only the deep culture method would be economically feasible.[66]

Civilian Distribution

In April 1944 an Office of Civilian Penicillin Distribution was set up in Chicago, Illinois, with John H. McDonnell as its director.[68] This unit controlled the distribution of such quantities of the drug as could be made available to a limited group of civilian hospitals. Initially, 1,000 hospitals were designated for receipt of penicillin as needed.[69,70] Gradually the number increased to 2,700 hospitals, using approximately 400,000 vials (100,000 Oxford units each) per month.[71]

By early 1945, the supply of penicillin was rising faster than demand for it. A total of 394 billion units were produced in January, 405 billion in

TABLE 9.2   Monthly U.S. Production of Penicillin, 1943–1945

Unit: billion units	
1943	
January to May, inclusive	0.400
June	0.425
July	0.762
August	0.906
September	1.787
October	2.872
November	4.846
December	9.194
Total	21.192
1944	
January	12.550
February	18.726
March	32.191
April	74.963
May	94.132
June	117.527
July	128.972
August	163.480
September	196.574
October	229.950
November	270.584
December	293.736
Total	1,663.385
1945	
January	394.113
February	405.156
March	460.958
April	510.960
May	615.071
June	646.817
July	616.897
August	636.51
September	586.37
October (est.)	620.00
November (est.)	660.00
December (est.)	700.00
Total (est.)	6,852.000

From: R. D. Coghill and R. S. Koch, "Penicillin a Wartime Accomplishment," *Chemical and Engineering News* 23, no. 24 (December 25, 1945): 2310.

February, and 460 billion in March (see table 9.2). On March 8, 1945, the WPB issued a release (cleared through the facilities of the Office of War Information) allowing these hospitals to purchase penicillin through distribution channels. Thus, penicillin became generally available for treatment of civilian patients.

PART III

The Years That Followed

10

The World Aroused

The story of penicillin and its development as a therapeutic agent relates mainly to events that transpired in England and the United States in the early 1940s. But, as already indicated, the development of chemotherapy as a science in large part evolved in Germany. The best chemotherapeutic drugs known to clinical medicine up to 1935 had emanated from the chemical laboratories of Germany, particularly those of the I. G. Farbenindustrie in Elberfeld. Even prontosil, from which all studies on the sulfonamides emanated, was discovered by a German chemist. It was to be expected, therefore, that penicillin would arouse interest among scientists at the I. G. Farbenindustrie and among clinicians throughout Europe.

In Germany and in Switzerland, where attempts to produce penicillin began as early as 1942, little was accomplished.[1,2] The instability of the substance and the small amounts produced in fermentation liquors thwarted investigators in these countries, as they had thwarted others. In Germany, in line with the approach so successfully used with the sulfonamides, the chemical nature of the inhibitor (that is, of penicillin) assumed importance, so that those assigned to its study were more skilled as chemists than as biologists. Emphasis was placed primarily on the chemical extraction, characterization, and possible synthesis of the compound.

Penicillin can now be synthesized, admittedly at considerable cost. But it was originally—and even today is primarily—derived from fermentation with appropriate strains of *Penicillia*. It was in those companies and those countries where microbiology as a science (and particularly fermentation technology) was highly developed that interest in penicillin production was most rapidly aroused. Holland, France, and Austria had the knowledge of fermentation technology and industrial microbiological processes to allow them to succeed even in the face of wartime shortages.

At the Yeast Factory in Holland

The German invasion of Holland occurred in May 1940—just three months before Florey and his associates in Oxford published their first report on penicillin. As a consequence, it was 1944 before information reached Gist-Brocades N.V. suggesting that penicillin production should be investigated. This company, the *Gistfabriek* (or "yeast factory," *Gist* meaning "yeast"), was prepared for the production of penicillin, for fermentation was the basis of everything the company had ever done.

The Netherlands' fermentation industries had been founded in Delft in 1860. From then on, the Gistfabriek had centered its attention on yeast fermentation. Initially its research was directed by the famed microbiologist Dr. M. W. Beijerinck, a professor at the Technical University of Delft. Later Beijerinck was succeeded by Prof. Albert Jan Kluyver, a microbial biochemist unsurpassed in the field of fermentation technology. The Gistfabriek research laboratories were designed specifically for the study of microorganisms detrimental to the growth of yeast cells and for the development of methods to combat growth of organisms harmful to yeast. The company had long used yeast in large fermentors for submerged fermentation of the organism.[3,4]

During the course of World War II and the occupation of the Netherlands, two companies in Holland attempted to produce antibiotics. One, the N.V. Koninklijke Pharmaceutische Fabrieken V. Brocades, Stheeman and Pharmacia, succeeded in producing a substance they called expansine (also known as patulin). This was produced by a strain of *Penicillium expansum* Westling but was too toxic for human use.[5,6] The other firm, Koninklijke Nederlandse Gist- and Spiritus-Fabriek N.V., was more successful. Unknown to the Germans, it succeeded in producing benzyl penicillin (penicillin G) by surface fermentation. The two companies eventually merged, became Gist-Brocades N.V., and together brought to the early production of penicillin in Holland the heritage of more than seventy years of fermentation experience.

Gist-Brocades first became involved in penicillin production when an English radio broadcast was picked up and "Flying Dutchman" pamphlets were dropped by British bombers. An issue of the *Klinische Wochenschrift* containing an article by Dr. M. Kiese on "Chemische Therapie mit Antibacterien Stoffen aus Niederen Pilzen und Bakterien" was obtained by Gistfabriek. At about the same time a report by Dr. A. Wettstein that described the results the Allies had achieved with penicillin also came to their attention. They already had available the necessary facilities and technical knowledge—in particular, knowledge of methods for mechanically mixing air and liquid in small fermentation units, a procedure essential for submerged growth of aerobic organisms.

Cultures of penicillia were obtained from the Central Bureau for Fungus Cultures, and by August 1944, they were producing a penicillinlike substance by surface fermentation in milk bottles. By 1945 pilot plant submerged growth fermentors with a capacity of 1,500 liters each were in operation. In April of that year, when American food parcels containing penicillin were dropped over the Ypenburg airfield between Delft and The Hague, they were able to establish that the antimicrobial substance they had been producing was the same as that which Florey, and by then many others, were producing under the name of penicillin. Until 1945 the substance produced in Holland was called bacinol.

In order to obtain large...amounts of this material, the fungus was cultivated on damp, sterilized wheat bran and malt culms which were extracted after the growth process. Starting from different liquid nutrients, cultures were...[grown] in thin layers in Roux flasks and in ordinary milk bottles. In order to follow day by day the formation of bacinol, a quantitative biological test [assay procedure] was developed, after which the concentration of a solution was expressed in Delft units. [The method of determining Oxford units was unknown in Holland at the time.]

This test was indispensable in attempts to concentrate and isolate the active substance, and...it was...[soon] found that bacinol could be extracted from an aqueous solution by means of ether and other solvents at a pH of 2 (in Britain it had been found that this was also possible with penicillin) and that it could be transferred again from these extraction agents to an aqueous buffer with a pH of about 7.[3]

Bacinol solutions were further purified by absorption on activated charcoal and elution with acetone, or by absorption in a column packed with chalk on silica gel. By June 1944, preparations of the calcium salt of bacinol with a purity greater than 50 percent were being produced.

When the liberation came in April 1945, British bombers dropped food and medicines, including a few ampuls of American-made penicillin on the Ypenburg airfield near Delft. With these ampuls of penicillin for comparison, it was established that the Delft product contained the same active substance as was being made in the United States. What is more, the purity of the Delft product was as great as that of the American product.

It was not until the war ended that it became known in Holland that Great Britain had been unable to produce adequate amounts of penicillin and had enlisted American help.

In October 1945 it became known in Delft that deep tank fermentation of penicillin was being used effectively in the United States and that the effect of corn steep liquor and phenylacetic acid—a precursor of penicillin—on penicillin yields was favorable. Phenylacetic acid had to be produced by the company itself (that is, by the Gistfabriek), and a feasible method had to be developed for its production. Thus, it was March 1946 before Gist-Brocades began to use phenylacetic acid as a precursor and initiated its first submerged fermentation runs. The ammonium salt of penicillin was produced at first by absorption onto a silica gel–chalk column, and the resultant salt had an activity of approximately 400 Oxford units per milligram. The first amorphous sodium benzyl penicillin (penicillin G) was produced commercially in the same year, and by 1948, Holland's total requirement for penicillin was being met by the Gistfabriek.[7]

In later years, larger fermentors were put into use for the manufacture of penicillin, and in 1949, the first crystalline sodium salt of penicillin became available for export from Holland.

Penicillin and the Occupation of Paris

We were at the end of 1942. France was surrounded everywhere with a thick wall which prevented news of the outer world...reaching...us. However, on December 31, our Scientific Department [at Rhône-Poulenc] received from Mr. Meyer, Director of our laboratories at St. Fons, a copy of a Swiss journal which had gotten into the Southern zone. This journal contained a "London Letter" which related the extraordinary results obtained in the treatment of war wounds with a new medicament, called Penicillin, an extract from a mold.... The word "Penicillin" did not awaken any echo in our minds. However...the facts seemed so definite that it appeared to us improper purely and simply to neglect them.[8]

On January 7, 1943, the then scientific director of Rhône-Poulenc, a pharmaceutical firm in France, visited Prof. F. Nitti at the Pasteur Institute in Paris to obtain a suitable penicillium culture for production of penicillin.[8,9] Professor Nitti had not heard of penicillin before, but since he had carried out many early studies on the sulfonamides, he was convinced of the merits of chemotherapy. His interest in penicillin was easily aroused.

In the beginning, experiments at the Pasteur Institute were discouraging, for several of the strains of penicillia they tested were not good penicillin producers. But in August 1943, Professor Nitti found, in a laboratory of the Institute, a subculture of Fleming's original strain of *Penicillium notatum*. By September, Rhône-Poulenc investigators were actively engaged in studies on cultivation of the organism and extraction of the penicillin produced by its growth.

On October 27 of the same year, 40 milligrams of crude penicillin were delivered to Professor Nitti for assay of its activity, which showed that the material contained 40 units of active penicillin per milligram. On December 6, Nitti reported data to the French Association of Microbiologists that fully confirmed the observations of Florey and the group in Oxford.[8]

From then on, penicillin production varied from month to month. Wartime needs were too important to permit its use for studies of academic interest only, and virtually all was used for clinical purposes. From October 1943 through December 1944, only a little more than 5 million units were produced (see table 10.1). Up to the time of the liberation, this was all the penicillin available in France.[8,9]

The production of penicillin during the "Occupation" was difficult. The "glass works" in France were not operating, gas and electrical supplies were often interrupted, and security alarms constantly drew people from their work. There

TABLE 10.1   Penicillin Production in France, 1943–1944

1943	October	1,600 Units
	November	4,262
1944	January	83,560
	February	191,000
	March	265,000
	April	165,000
	May	280,520
	June	543,000
	July	1,608,000
	August	516,000
	September	82,125
	October	547,125
	November	55,000
	December	765,000
	Total	5,107,192 Units

Source: Personal communication from Dr. Georges Werner, and the files of Rhône-Poulenc (1978).

was no standard penicillin with which to compare the material produced, and no quantitative or reliable procedures for its biological assay or for determination of its toxicity.

In addition, "there were the Germans. It was known, in devious ways, that they were not indifferent...to penicillin.... We wished to avoid their concerning themselves with our investigations and we did our utmost to avoid attracting their attention. And we succeeded. At no time did the invaders interfere with our investigations in any way..." At the Pasteur Institute, on the other hand, the Occupation Authorities demanded...delivery of a culture of the *Penicillium notatum* strain. A culture of *Penicillium* was given to them, but it is scarcely probable that it ever furnished them much penicillin.[9]

The first samples of American-made penicillin reached France with the Allied troops following the liberation. From these, it was learned that the unit of potency used within Rhône-Poulenc had been very similar to the English Oxford unit.

After the liberation, in late 1944, the Military Health Department in France set up a small surface growth penicillin factory to supplement the amount supplied by the Allies. Aided by the Pasteur Institute, it undertook to cultivate the penicillin-producing organism, while Rhône-Poulenc extracted the drug from the fermentation liquor. By late April 1945, penicillin was being produced by fermentation in the plant of the Military Health Department. The extraction plant at Rhône-Poulenc had been in operation since January, and between then and April, penicillin was recovered by Rhône-Poulenc from the urine of wounded American servicemen being treated in the hospitals in the Paris district. The penicillin recovered from the Americans was used to treat the wounded of the French army. About 100,000 units (100 doses) were obtained routinely from 300 liters of urine.[10]

From April on, production increased steadily as did the flow of information from the United States and elsewhere. In July 1945, production of penicillin in France reached a level of more than 300 million Oxford units, and in December, the French government authorized Rhône-Poulenc to negotiate with an American firm to set up a modern submerged growth penicillin plant.

At Biochemie in Austria

Höchholtingen Castle in Kündl was built in 1495; a brewery was located on the grounds continuously from 1658 until 1945. After 1927, the brewery belonged to Oesterreichischer Brau AH in Linz, a company that closed in 1945 for want of raw materials. The empty brewery was ideal for penicillin production and Biochemie GmbH was established for this purpose. The new firm developed its own process and the plant opened in May 1948.

That date, of course, is late in the history of penicillin development. Nevertheless, the company's place in the penicillin story is important. Biochemie at first used brewing yeast autolysate in lieu of corn steep liquor for production of penicillin. Crude lactose from Reutte, Tannheim, and from Salzburg was used rather than purified lactose. Butanol was produced by Biochemie, by a fermentation process, for use in the extraction. A freeze-drying process was employed, which yielded an amorphous yellow-brown powder.

By 1951, Biochemie was able to meet a large portion of Austria's penicillin requirements, but it became apparent that the plant must be enlarged. During the enlargement, production continued in the old plant, but under difficult circumstances, for the new plant was built in the same rooms used by the old plant. Contamination—growth of undesirable bacteria in the penicillin culture medium—led to rapid destruction of the already formed penicillin. During the course of research on methods to combat this contamination, Dr. Ernst Brandl, a graduate student still working on a dissertation for his Doctor of Philosophy degree, noted that, through a reaction of a disinfectant beta-phenoxyethanol (which he had added to the fermentation broth), a previously unknown substance—phenoxymethyl penicillin—was formed. This crystallized in a free-acid form and proved to be stable in acid. This acid-stable penicillin V, as it was designated, opened the door for the use of penicillin orally which until then had not been possible because of penicillin's instability in gastric juice.

Subsequent research at Biochemie, directed toward the search for new antimicrobial and synthetic substances, led to the realization that the penicillin V side-chain could be split off enzymatically and 6-aminopenicillanic acid could be isolated. The preparation of some of the earliest new biosynthetic penicillins followed.[2,11,12]

In Germany

At the close of World War II, five executives from the U.S. pharmaceutical industry visited Germany to obtain information for the U.S. Department of Commerce.[13] They found that the I. G. Farbenindustrie had attempted production of penicillin both at its Elberfeld plant and at Hoechst but with little success. At Hoechst, research on penicillin had begun in 1942, and by 1943 the company was producing small amounts of the drug. Initially round-bottomed flasks were used; later, flat sheet-metal "vats" came into use, the vats being rocked mechanically to increase the rate of growth. By 1944, yields equivalent to 50 Oxford units per milliliter of fermentation liquor were obtained, and Hoechst penicillin in the form of a yellow amorphous powder was available for use in a limited number of patients.

Both the surface growth and submerged growth processes for manufacture of penicillin were in operation at Hoechst in 1945. The Czapek-Dox medium was used as base, with lactose (in lieu of glucose) and yeast extract added. Optimum pH, they had determined, was 6.4. In their experience, maize, wheat, soya, peptone, peanut meal, and oats had offered no special advantage. In the surface growth process, growth of the penicillium reached a maximum at about nine to fourteen days. Ample aeration was considered essential. The active penicillin was extracted in butyl alcohol.

In January 1945, Hoechst planned to expand its penicillin-producing facilities. This was not accomplished, however, for in March of that year the Americans occupied the Hoechst factory and production temporarily ceased. The plant, however, reopened within the same month, and by April, Hoechst was producing penicillin (as the calcium or sodium salt) for parenteral use, as a powder (of the calcium salt) for topical use, and incorporated in bandages. All preparations, moreover, were catalase-tested to ensure against contamination by notatin (expansine or penicillin B).

Output by the surface growth technique reached 30 million units per month by November of the next year. However, Brig. Gen. William Draper, who later became under secretary of defense, had proposed earlier in 1945 that the United States build a modern and properly equipped penicillin-manufacturing plant in Germany to help combat some of the diseases prevalent there and to help revive the German economy. The already available Hoechst plant was used, and Merck & Co. of Rahway, New Jersey, supervised construction and conversion to an up-to-date plant for submerged fermentation of penicillin. The plant opened in mid-1950.

The United States had again shared its penicillin production know-how with another country—this time with a member of the Axis, its wartime enemy.[2,14,15]

In Japan

In Japan research on penicillin was initiated on February 1, 1944. Dr. Katsuhiko Inagaki, a surgeon, received on December 21, 1943, a copy of an article by Manfred Kiese[15] that had aroused the interest of Delft investigators in penicillin. Kiese's article was translated from German into Japanese by Dr. Hamao Umezawa. A penicillin committee was formed and held its first meeting on February 1. Dr. Inagaki quickly became the prime promoter of research on penicillin in Japan, and Dr. Hamao Umezawa became the the country's foremost researcher on the biological aspects of penicillin. His brother, Dr. Sumio Umezawa, assumed a similarly prominent position, devoting his attention to studies of the chemical nature of penicillin.

It was December 21, 1943 when Dr. Katsuhiko Inagaki, a surgeon...who became a promoter of the penicillin research during the war, visited the Ministry of Education, where he met Willi Nagai of the Research Section, a son of the famous Nagayoshi Nagai who discovered ephedrine. He was given a few copies of medical journals sent from Germany by submarine. Inagaki was particularly interested in a review article written by Manfred Kiese, of the Pharmacological Laboratory of Berlin University, on chemotherapy with antibacterial substances obtained from lower fungi and bacteria. In this article, published in the August 7, 1943 issue of *Klinische Wochenschrift*, Kiese abstracted papers of the Oxford group, Drs. Florey, Chain, Abraham, and others, on penicillin production, chemistry, activity, pharmacology and clinical effects, which were reported in *Lancet*, *Nature*, and *Biochemical Journal* during the period 1940 to 1943. Papers on gramicidin and tyrocidin, actinomycin as well as notatin were reviewed briefly; and other antibacterial substances such as spinulosin, fumigatin, citrinin, kojic acid, mycophenolic acid and griseofulvin were also mentioned in this review.[16]

More than thirty years later, Drs. Hamao and Sumio Umezawa remained leaders in Japan in research on antimicrobial drugs.

An official notice, dated January 27, 1944, was sent from Dr. T. Okada, a chief of the Hygienic Section of the Ministry, to Dr. S. Uebayashi, an executive secretary of the Military Medical College. It was requested in the notice that the Military Medical College should control and promote the research on purification and synthesis of penicillin and other antibacterial substances, to apply them to military medicine practically, and that a plan and budget should be presented. It was also stated that a person in charge of research control should be appointed among the staffs of the college; that non-military researchers should work in their own laboratories, mainly on basic research as non-regular staff members of the military;

that meetings should be held, from time to time, to report the progress and to promote the research which was expected to be completed in August, 1944; and that an additional budget of about 150,000 yen was pending. This was the usual procedure of the Japanese Army. In fact, the project was approved by the War Minister, and the Collegemaster, Dr. Y. Miki, a surgeon lieutenant-general, became...chairman of the committee. Inagaki became...promoter of the penicillin project.

January 27 was...a momentous day for Japanese penicillin. A cable... [from] Mr. Y. Imai from Buenos Aires, Argentina, was printed in the *Asahi*, one of the most well-read daily newspapers. It was a long article.... We were informed, by this news, how penicillin was manufactured, and how it was used effectively in the treatment of infections. It was also reported that large-scale production was being carried out in Peoria, Illinois, at the Northern Regional Research Laboratory of the U.S. Department of Agriculture.[17]

Progress made by the penicillin committee in Japan in the following months was recorded by Y. Yagisawa in 1980.[17] Production by the Banyu Pharmaceutical Company began in January 1945, when Japan was being heavily bombed. In August, the United States dropped the first atomic bombs on Hiroshima and Nagasaki, Russia declared war on Japan and invaded Manchuria, and Japan announced its surrender on August 14.

Japan had started its investigation of penicillin at a very late date, and because of events related to World War II, its course was stormy. Nevertheless, a powerful and highly productive scientific program arose, encompassing not only penicillin but many other antimicrobial drugs. Now, almost forty years later, Japan ranks second to none in its expertise in the field of antimicrobial agents.

In Canada

In May 1941, Howard Florey visited the Connaught Laboratories, a firm that was an integral part of the University of Toronto in Canada. The university's School of Hygiene was staffed in large part by members of the Connaught Laboratories, and together they carried on an integrated program of research, teaching, and product development. Following Florey's visit, work on penicillin production was started by Dr. Philip Greey in the Department of Pathology. Later Drs. C. C. Lucas and S. F. MacDonald of C. H. Best's department joined the project. They were quite successful, and by August 1943 the Connaught Laboratories had adequate facilities for what they considered to be large-scale production.

At the request of the director general of the army's medical services in Ottawa, the Connaught Laboratories undertook the manufacture of penicillin. An added source of supply was needed, and Ayerst, McKenna and Harrison of Montreal was invited to participate. The dominion government supplied the funds needed by the Connaught Laboratories for remodeling a section of a building acquired for this work, while the university undertook for Connaught the preparation at cost plus a 5 percent overhead charge. The university purchased for the purpose the old Knox College property. On October 1, 1943, Canadian investigators requested that the Connaught Laboratories be included in the U.S. War Production Board's group of seventeen manufacturers so that they might benefit from conference discussions.

By December 1945, three companies in Canada were producing approximately 20 billion units of penicillin monthly, an amount sufficient to meet all Canadian needs.[17,18]

Penicillin Worldwide

Penicillin production in the United Kingdom by December 1945 had reached a level of 30 billion units per month produced by seven companies; it was anticipated that soon their output would reach 200 billion units monthly. In Australia, a government-owned plant was producing more than 10 billion units per month, and in Mexico small amounts were becoming available. Elsewhere the drug was still nonexistent. Our enemies in World War II had developed no antibiotic drugs, and the mortality among their wounded had remained high—far higher than among the wounded of the United States or British armed forces.

The United States initiated export of penicillin in June 1944, but until April of the next year, only producers, not distributors, of penicillin were permitted to export because of the limited supplies and controlled distribution procedures in the countries of import. When penicillin became more readily available to civilians in the United States in March 1945,

penicillin control committees were no longer required in other countries, and anyone could export penicillin from the United States, provided that he had the penicillin or had a firm commitment from a supplier, and that the license application was received in time to permit it to be charged against Foreign Economic Administration monthly quotas established by the WPB Chemicals Bureau Requirements Committee on Drugs.[19]

By December, the export of 200 billion units per month equaled domestic consumption. The export demand continued to rise for some time thereafter. In the first seven months of 1945, 721.6 billion units were exported—400 billion of that number in July and August alone. The United States

shared penicillin with other countries for military and civilian use on the basis of an equitable division of the available supply and at comparable cost, thereby contributing to the improvement of public health throughout the world.[19]

As Robert Coghill of the Northern Regional Research Laboratory and Roy Koch of the Civilian Production Administration (formerly the War Production Board) rightly commented toward the end of 1945, penicillin had done its share "to establish a better understanding among nations, which should aid in the establishment of world peace."[19]

By the end of 1945, production in the United Kingdom averaged 30 billion units of penicillin per month, with seven firms producing and a potential of 200 billion units per month within the next year. In Australia, a government-owned firm had started operation in 1944 and was producing approximately 10 billion units monthly by the end of 1945. Canada's production approached 20 billion units, sufficient for the entire country; and Mexico was producing smaller (but significant) amounts. Some other countries had plants in early stages of construction but none produced significant amounts at that time.

In recognition of Fleming's contribution to science and mankind through his discovery of penicillin, the American penicillin-producing industry presented him with a trust fund of $100,000 to finance research conducted under his supervision at the St. Mary's Hospital Medical School, University of London. This was a truly large research grant in 1945; the University of Pennsylvania where Dr. A. N. Richards, chairman of the CMR, had been affiliated as professor of pharmacology became administrator of the fund.[19]

11

New Penicillins Introduced

In 1970, twenty-five years after D-Day, Dr. Albert Elder (penicillin coordinator for the War Production Board and the Office of Scientific Research and Development in 1943) chronicled the manufacturing achievements that had led to penicillin's successful production in so short a period of time.[1] In his judgment, the wartime challenge to produce massive amounts of the drug in the face of inadequate technical knowledge stimulated thinking, which led to the success. But the real key, he claimed, was the complete and free exchange of information among scientists and throughout the pharmaceutical industry.

In order to facilitate the rapid solution of technical problems related to penicillin production on a commercial scale, the WPB agreed to waive the application of U.S. antitrust laws to collaborating activities of drug manufacturers in the production of penicillin. "The action was taken at the request of the Office of Scientific Research and Development...in a certificate to the Attorney-General dated December 7 [1943]." Suspension of the antitrust law controls applied only to the exchange of technical information related to penicillin production, not to the production of other drugs.[2,3]

The free exchange of information made possible by the lifting of the U.S. antitrust law controls undoubtedly sped mass production during 1944–45. Later it may have led to increased competition among firms that might not otherwise have undertaken to manufacture the drug commercially. By December 1943, however, the day when penicillin would be available in quantity was rapidly approaching. Production was under way or due to start in twenty-one firms and it was anticipated that it would reach such proportions by mid-1944 that all military and pressing civilian requirements could be met.[2]

The Nobel Prize in Physiology or Medicine

Alexander Fleming, Howard Walter Florey, and Ernst Chain received the Nobel Prize in physiology or medicine in 1945 "for the discovery of penicillin and its curative effect in various infectious diseases." Previously, in June 1944, Fleming and Florey had been included in the first birthday honors list; each had been created a knight bachelor at that time.[4] Later, in 1965, Chain was so honored, and in 1980 Abraham was also created a knight bachelor. The Nobel Prize was presented to Fleming, Florey, and Chain by Professor Göran Liljestrand, member of the staff of professors of the Royal Karolinska Institute in Stockholm, Sweden.[5] Liljestrand, as chairman, steered the Nobel prize committee wisely. Lord William

PLATE 11.1. *Howard Florey, Ernst Chain, and Alexander Fleming at the presentation of the Nobel Prize by the king of Sweden. (Reproduced with permission from: A. Maurois,* The Life of Sir Alexander Fleming *[New York: E. P. Dutton & Co., 1959], opp. p. 224.)*

Beaverbrook, British statesman, newspaper owner, and a member of Churchill's wartime cabinet, had exerted great pressure through the press to have the award given to Alexander Fleming only, whereas Prof. John Fulton in the United States pressed hard for it to be awarded to Howard Florey only. Strangely, Fulton never recognized that the discovery of the substance was basic to its ultimate development.[6] The prize was awarded to all three men in December 1945.

In accepting the Nobel Prize for the discovery of penicillin, a discovery for which he alone was responsible, Fleming described the circumstances surrounding his discovery and gave credit to his two "co-participators" in the award by remarking that Chain and Florey had succeeded in concentrating penicillin to such a degree that "now...it is active beyond the wildest dreams I could possibly have had in [the] early days [after its discovery]."[7] Florey spoke of the possible role of antimicrobial substances in nature, a subject of special interest to him and one that had led Chain and himself to the study of penicillin.[8] Chain, in turn, presented a classic lecture on all that had been done to that time and all that was known concerning the chemical structure of penicillin.[9]

The Chemical Structure of Penicillin

From the beginning, it was believed that methods for the chemical synthesis of penicillin would be forthcoming at an early date. This was the fear, in fact, of those equipping and developing expensive fermentation plants. But between 1941 and 1944, it became clear that the production of penicillin could perhaps be accomplished more easily by fermentation than by chemical synthesis. By the end of that period, experience had indicated that this probably would be the case. Supplies of penicillin increased rapidly, and its cost began to fall. The need for methods allowing chemical synthesis had begun to diminish.

Studies on the purification and structure of penicillin were initiated at Oxford in 1941 by Dr. Ernst Chain and Dr. Edward Abraham. Late in 1942, Dr. W. Baker (professor of organic chemistry at Bristol) and Sir Robert Robinson joined them in an attack on the problem. Together they elucidated the structure of the penicillins and synthesized some of their degradation products (see figure 11.1).[9,10]

The penicillins, they observed, are organic acids, readily soluble in different organic solvents, but insoluble or only sparingly soluble in hydrocarbons. They are stable in water (in a pH range between 5 and 8)

BENZYL PENICILLIN

RHCH₂CO ¦ NHCH ── CH $\underset{\underset{\text{(a)}}{\text{CO ── N}}}{\overset{\overset{\text{S}}{\diagup}}{}}$ C $\overset{\text{CH}_3}{\underset{\text{CH}_3}{}}$ CHCOOH

(b)   (a)

a. Site of penicillinase (β-lactamase) action
b. Site of amidase action

THE PENICILLIN NUCLEUS
(6 - Aminopenicillanic Acid)

H₂N ── S ── CH₃ / CH₃
O ═ ── N ── COOH

(a) From fermentation   R =   ⬡-CH₂-   Penicillin G
    of *Penicillium*

                           R =   HO-⬡-CH₂-   Penicillin X

                           R =   CH₃-(CH₂)₆-   Penicillin K

(b) From fermentation
    with synthetic   R =   ⬡-O-CH₂-   Penicillin V
    precursors

(c) From fermentation   R =   H₂N-CH-(CH₂)₃- Penicillin N
    of *Cephalosporium* |
                                   COOH

Adapted from: Lancini, G., and Parenti, F: Antibiotics, An Integrated View, 1982, page 96:
Springer-Verlag, New York, Heidelberg, Berlin; and Garrod, L.P., et al.: Antibiotic Chemotherapy,
3rd Edition, 1972, page 54: The Williams & Wilkins Company, Baltimore.

SOME BIOSYNTHETIC PENICILLINS

NAME	SIDE CHAIN	IMPORTANT PROPERTIES
Benzyl penicillin 6-phenylacetamido penicillanic acid	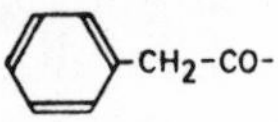 $-CH_2-CO-$ (benzene ring)	
Phenoxymethyl penicillin Penicillin V 6-phenoxyacetamido penicillanic acid	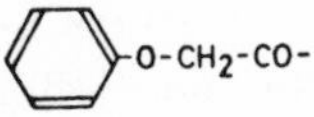 $-O-CH_2-CO-$ (benzene ring)	**Acid-resistant**
Phenethicillin DL-6-(α-phenoxypropionamido) penicillanic acid	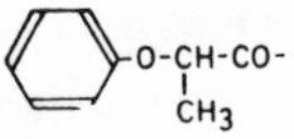 $-O-CH-CO-$, CH_3 (benzene ring)	**Acid-resistant**
Propicillin DL-6-(α-phenoxy-n-butyramido) penicillanic acid	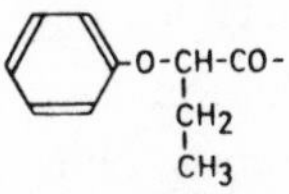 $-O-CH-CO-$, CH_2, CH_3 (benzene ring)	**Acid-resistant**
Ampicillin 6-(D(-)-α-aminophenylacetamido) penicillanic acid	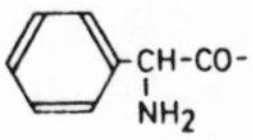 $-CH-CO-$, NH_2 (benzene ring)	**Active against Gram-negative bacilli Acid-resistant**
Carbenicillin Disodium α-carboxy-benzyl- penicillin	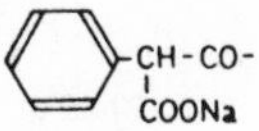 $-CH-CO-$, $COONa$ (benzene ring)	**Active against Gram-negative bacilli**
Methicillin 6-(2,6 dimethoxybenzamido) penicillanic acid	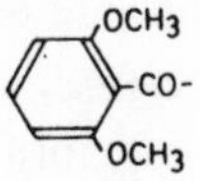 OCH_3, $-CO-$, OCH_3 (benzene ring)	**Penicillinase-resistant**
Cloxacillin 6-(5-Methyl-3-orthochlorophenyl- isoxazole-4-carboxyamido) penicillanic acid	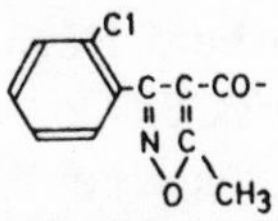 Cl, $-C-C-CO-$, N C, O CH_3 (benzene ring)	**Penicillinase-resistant Acid-resistant**

Adapted from: Garrod, L.P., et al.: Antibiotic Chemotherapy, 3rd Edition, 1972, pages 54, 70–71; 5th Edition, 1981, pages 58, 95; The Williams & Wilkens Company, Baltimore.

FIGURE 11.1. The structure of penicillin and side chains of some biosynthetic penicillins.

only in the form of their salts and rapidly lose biological activity in aqueous solutions of higher acidity or alkalinity. The penicillins are also inactivated by many other reagents in addition to acid and alkali. For example, they are inactivated by some heavy metal ions, by primary alcohols and amines, thiols, aldehydic or ketonic reagents, oxidizing reagents and a specific enzyme, penicillinase, which is produced by some penicillin-resistant strains of bacteria.

Before long, others began to attack the problem. Among them were Dr. A. H. Cook and Prof. Sir Ian Heilbron at the Imperial College of Science; Dr. S. Smith and his associates at Burroughs Wellcome; investigators at the Imperial Chemical Industries; and Sir Robert Robinson and the group at Glaxo Laboratories. Simultaneously, chemists in the United States began an intensive study of the structure of penicillin with chemical synthesis as an ultimate goal.

During the first part of 1943, studies conducted primarily in Great Britain established that, in addition to penicillamine, 2-pentenylpenillic acid and 2-pentenylpenillamine could be isolated as conversion products of the crude penicillin then available. Evidence was obtained at the Imperial College of Science in London to show that penicillin, on degradation, yielded a product that appeared to be an amino acid, and the Wellcome Research Laboratories reported the conversion of penicillin in acid solution to a crystalline product, which was designated penicillic acid.[9-11]

In mid-1943, MacPhillamy, Wintersteiner, and Alicino at the Squibb Institute in New Brunswick, New Jersey, succeeded in crystallizing the sodium salt of benzyl penicillin.[12] This facilitated confirmation of the presence of sulfur in the penicillin molecule and in various penicillin derivatives including penicillamine (which proved to be beta, beta-dimethylcysteine). It opened the door, moreover, to recognition of differences in the penicillins then produced in the United States and England. Soon Frank Stodola at the Northern Regional Research Laboratories in Peoria and investigators at Merck & Co., the Imperial Chemical Industries, and Oxford produced evidence confirming differences between the British and the American penicillins. In particular, it became apparent that the American preparations yielded phenylacetic acid on hydrolysis, whereas British preparations yielded 2-hexenoic acid under similar conditions.[11]

Four "naturally" occurring penicillins were recognized. All had qualitatively similar biological properties but they differed in chemical composition; all contained a common nucleus but differed in the structure of their side chains; each was isolated in the form of its crystalline sodium

PLATE 11.2.   Dr. Oskar Wintersteiner in whose laboratories at the Squibb Institute for Medical Research benzyl penicillin was first crystallized by Dr. H. B. MacPhillamy and the empirical formula $C_{16}H_{17}O_4N_2SNa$ (sodium benzyl penicillinate) was established.

PLATE 11.3. Dr. Joseph F. Alicino of the Squibb Institute for Medical Research, who first demonstrated the presence of sulfur in the penicillin molecule and later developed the widely used iodometric assay of penicillin.

salt. In England, they were designated penicillin I, II, III, or IV, according to the historical order in which they were discovered. In the United States, they were designated penicillin F, G, X, and K.[11,13] Later it became accepted practice to use chemical designations, for example:

	British Designation	American Designation
2-Pentenyl penicillin	Penicillin I	Penicillin F
Benzyl penicillin	Penicillin II	Penicillin G
Para-hydroxybenzyl penicillin	Penicillin III	Penicillin X
Heptyl penicillin	Penicillin IV	Penicillin K

By the addition of appropriate precursors to culture media, other naturally occurring penicillins were soon produced. However, Only phenoxymethyl penicillin (penicillin V), which is obtained when phenoxyacetic acid is used as the precursor, assumed importance. First described in 1948, it was rediscovered in Austria in 1953 (see chapter 10), where it was shown to be stable in the presence of gastric juice and effective when administered orally.[11-13]

Now it is recognized that "penicillin" is not a single entity. Rather, it is a family of antibiotics, that is, a family of penicillins, all of which contain a beta-lactam ring. As a class, the beta-lactam antibiotics are probably the most powerful antibacterial agents discovered to date (see chapter 1).

Using fermentation methods in an extended study of penicillin precursors, Behrens et al. at Eli Lilly & Co. succeeded in making a phenoxymethyl side chain that conferred acid stability upon the penicillin molecule. It was already known that p-hydroxyphenylacetic acid and other derivatives of acetic acid could be incorporated biosynthetically by the mold into the penicillin molecule. Several penicillins with substituted acetic acid side chains and with the antibacterial spectrum of penicillin G were thereby produced, in particular phenoxymethyl penicillin (penicillin V).[14]

The studies on the structure and synthesis of penicillin carried out by both the British and the Americans between 1941 and 1945 were described in detail in a remarkable monograph written and edited at the close of World War II. It encompassed all that had been accomplished under the auspices of the OSRD and the British Medical Research Council.[11,12]

The British and American chemists worked independently of each other until May 1944, by which time penicillin was being produced commercially in massive amounts by submerged fermentation. During the next eighteen months, however, there was free exchange of information on all chemical studies.

By the end of 1945, Dorothy Crowfoot and Dr. E. R. Low, in the Department of Crystallography at Oxford, and Dr. E. R. Holiday at the London Hospital had shown that much could be learned from small amounts of material, even crude material. Obtaining first crystallographic data on crystalline degradation products of penicillin and information on the ultraviolet absorption spectra of such substances, they later produced conclusive evidence of the beta-lactam structure of penicillin by such infrared absorption and crystallographic studies of the rubidium and potassium salts of the anitbiotic.[11,15]

Many were involved, up until late 1945, in these studies on the structure and synthesis of penicillin. As military need for the drug diminished, however, and supplies produced by fermentation increased,

PLATE 11.4. Dr. John C. Sheehan, professor of chemistry emeritus, Massachusetts Institute of Technology, who achieved the first total general synthesis of the penicillins.

most investigators went on to other projects. Only Dr. John C. Sheehan (originally affiliated with Merck & Co., later at the Massachusetts Institute of Technology where his studies were partially supported by Bristol Laboratories) continued the pursuit. He was successful, but only after many years.

The significance of penicillamine as an integral part of the penicillin molecule had been obvious from the start, and this penicillin degradation product (later shown to be beta, beta-dimethyl cysteine) served as a starting point in many studies. Small amounts of benzyl penicillin were produced by condensation of penicillamine with 2-benzyl-4-methoxy-

PLATE 11.5.   Crystalline sodium benzyl penicillinate, or penicillin G. (Courtesy of Dr. H. B. MacPhillamy, the Squibb Institute for Medical Research.)

methylene-5(4)-oxazolone, but this failed to offer a practical means of synthesis. Sheehan described the synthesis of 5-phenyl penicillin and described also two other beta-lactam syntheses yielding substances with the chemical and physical properties of natural penicillins. But these all lacked antibacterial activity. In 1950, Sakaguchi and Murao in Japan suggested that benzyl penicillin could be deacylated, and three years later, Kato, also in Japan, suggested that the intact beta-lactam nucleus could thereby be liberated. But he failed to isolate the penicillin nucleus.[16,17]

John Sheehan meanwhile continued his attempts at total chemical synthesis. He succeeded in making a pure sulphonyl analogue of benzyl penicillin (benzene-sulphonamido-penicillanic acid) which possessed some biological activity. But the obstacle to total chemical synthesis continued to be the cyclization of penicilloic acid to close the lactam ring of the nucleus. Not until 1956–57 did Sheehan show that N-substituted carbodiimide reagents would effect cyclization to form penicillin V from the corresponding penicilloic acid. He demonstrated then the presence of a key intermediate penicillinoate that allowed the synthesis of further derivatives.[18,19]

The synthesis of penicillin derivatives by the above route was difficult; yields were low and the process was costly. Penicillin V was the only useful derivative made by the procedure, and even this was easier to produce by fermentation.

Tailor-Made Penicillins

Penicillins are N-acylated derivatives of an amino acid that has been designated 6-aminopenicillanic acid (6-APA). The 6-APA nucleus is a fused beta-lactam-thiazolidine bicyclic system. The penicillins act by interfering with a specific cross-linking reaction that is essential for formation of bacterial cell walls. Each penicillin owes its biological activity to the side chain attached to a "nucleus" which is the same in all penicillins.

Total synthesis of the penicillin molecule by chemical means, as indicated above, was not accomplished in time to be useful in the development of penicillin as a chemotherapeutic drug, nor was it ever economically feasible. Reliance was placed on biosynthetic methods. One approach, used initially, was to add to the fermentation liquors substances that would provide a source of the required side chain and thereby act as precursors. Otto Behrens and his associates at Eli Lilly & Co. in Indianapolis in this way showed a remarkable degree of enhancement of penicillin

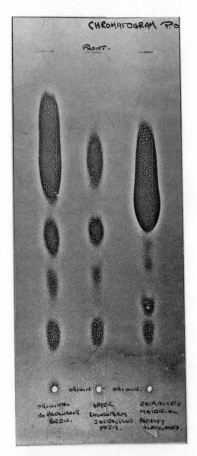

PLATE 11.6.   One of the first chromatograms showing the production of 6-APA component of penicillin V, Beecham Pharmaceuticals. (Courtesy of Dr. F. R. Batchelor.)

yields in fermentation liquors and established the potential of this approach. Gradually it became apparent that removal of the side chain from benzyl penicillin provided a convenient "skeleton" or starting point for the synthesis of new penicillins. For total synthesis of the penicillins, however, it was necessary to establish the nature of the "nucleus" and to synthesize it. After more than a dozen years of research on the problem, this was accomplished by Dr. John C. Sheehan at the Massachusetts Institute of Technology in Boston. Preparation of the penicillin nucleus (6-aminopenicillanic acid) was achieved, and total synthesis of penicillin was accomplished also.[18]

225

PLATE 11.7.   Dr. Otto K. Behrens, chemist at Eli Lilly & Co., who during the 1940s conducted a large-scale study of penicillin degradation products, metabolic products, intermediates, and similar substances to establish if any would stimulate the production of penicillin by penicillia, by acting as precursors.

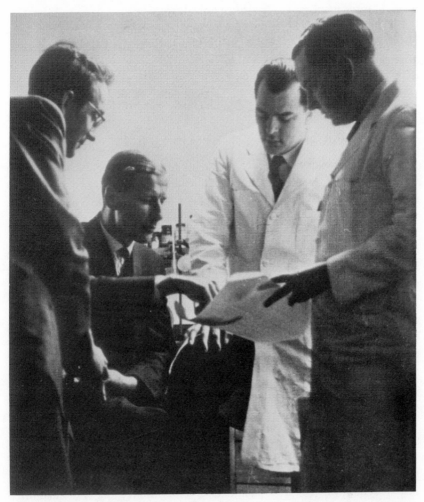

PLATE 11.8.   The scientists who isolated 6-APA, leading to the production of the new penicillins: (from left) *F. P. Doyle, G. N. Rolinson, F. R. Batchelor, and J. H. C. Naylor. (Reproduced from permission from H. G. Lazell,* From Pills to Penicillin—The Beecham Story *[London: William Heinemann, 1975], opp. p. 69.)*

The techniques used by Sheehan were masterful, but they were not suited to commercial use. It remained for those skilled in the art of fermentation to provide a more practical process.

On August 2, 1957, Frank Batchelor, Frank Doyle, John Nayler, and George Rolinson of the Beecham Laboratories in England filed a patent application on a process for isolation of 6-aminopenicillanic acid or a salt thereof from fermentation liquors. The patent was issued on June 21, 1960.[20-22]

The Beecham Story

Beecham Laboratories had been trying for some time to stabilize penicillin G for oral use. The introduction of penicillin V in 1954, however, had led researchers there to abandon their attempts with penicillin G and to seek a new approach.[22]

Sir Charles Dodds, a senior consultant to Beecham, urged that they obtain the advice and talents of Ernst Chain who by then had moved to Rome, Italy, and had constructed a small pilot plant for antibiotics research at the Istituto Superiore di Sanità. Encouraged by Sir Charles, they approached Professor Chain who suggested studies of (a) molecular modifications leading to penicillinase-stable and broad spectrum penicillins, and (b) new mutants of the mold. Dr. George N. Rolinson, a microbiologist, and F. Batchelor, a biochemist, were assigned to the project and proceeded to Rome where they could work under the guidance of Ernst Chain.

Rolinson and Batchelor remained in Rome for some months in 1956, during which time Beecham Laboratories optimistically constructed new microbiological laboratories and a plant at Brockham Park in England.

Batchelor and Rolinson at Rome began work on the biosynthesis of p-aminobenzyl penicillin, a starting point suggested by Chain. Collaborating with them were Drs. A. Ballio and Dentice di Accadia of the institute's staff. They hoped that the reactive NH_2 group on this derivative would provide a point of attachment for additional side chains. By adding p-aminophenylacetic acid as precursor to the fermentation medium, the p-aminobenzyl derivative was produced (to the extent of about 40 percent of the total penicillin activity). Initiated first in flasks, the fermentation was soon transferred to 50-liter vats, and by October 1956 (within only nine months after start of the project), a significant amount of the p-aminobenzyl derivative was available in an 80 percent pure form.

228

PLATE 11.9.   *The first penicillin plant installed at Worthing, England, Beecham Laboratories. (From H. G. Lazell,* From Pills to Penicillin—The Beecham Story *[London: William Heinemann, 1975], opp. p. 69.)*

Prior to 1952, a large number of penicillins had been made by modification of the penicillin molecule through the addition of appropriate precursors to the fermentation medium (see the Behrens study described above). It was known that chemical and microbiological assays on the products of fermentation were not always in agreement, and in Rome, it was noted that the difference was most marked when the appropriate side-chain precursor was omitted. This led to identification of the penicillin "nucleus," 6-aminopenicillanic acid, in the fermentation liquors. It was obvious that 6-aminopenicillanic acid was itself produced as a natural product. Chain had sensed this from the beginning, saying repeatedly, "It has to be in nature." When the substance was detected, therefore, no one was surprised. It was the last piece in the puzzle.

The penicillin "nucleus," 6-APA, had the properties of penicillin in that it reacted with hydroxylamine and was destroyed by penicillinase. But it lacked the in vitro antibacterial activity of the penicillins. Acetylation of the inactive substance in May 1957, by the addition of phenylacetyl chloride, yielded benzyl penicillin. By the use of paper chromatography and bioassay of the chromatography strips, initial results were confirmed, and it seemed logical to assume that if 6-APA could be isolated the potential of producing new penicillins by side-chain substitution at the 6-position would be enormous.

Unstable to heat and unstable in solution, 6-APA was amphoteric and hygroscopic. The process of extracting it from culture fluids was not simple, but its isolation was accomplished in 1957. Following a lead from Japan,[17] the Beecham group[23] then demonstrated that *Streptomyces lavendulae* and some other molds produce an enzyme (an extracellular amidase) capable of yielding 6-APA by deacylation of preformed penicillins.

The first new and semisynthetic penicillin, 6-(*L*-phenoxymethyl-propionamido)-penicillanic acid, or *L*-phenoxymethyl penicillin (officially designated phenethicillin, although informally known as penicillin V), was marketed in late 1959. Developed partly by Beecham Laboratories and in large part by Bristol Laboratories, it was the first in a short series of acid-stable penicillins that could be administered effectively by the oral route.[24] Other penicillin derivatives, some stable in the presence of penicillinase, some with broader spectrums of activity than that of benzyl penicillin (i.e., penicillin G) followed in quick succession.[25-29] Made in part by chemical, in part by biological means, they quickly established the principal of semisynthetic penicillins. Beecham, later with help from the

Bristol Laboratories,[25] had paved the way for far-reaching developments in the fields of antimicrobial drugs and infection control, the significance of which became increasingly apparent during the succeeding twenty or more years.

Methicillin, active against penicillinase-producing staphylococci, was introduced in 1960. Ampicillin, the first broad-spectrum penicillin, came to the fore in 1961. Cloxacillin, orally active against penicillin-resistant staphylococci, was introduced in 1963, and 1967 saw the introduction of carbenicillin, a penicillin active against strains of the Pseudomonas species of microorganisms.

Penicillins are beta-lactam antibiotics of which there are two major categories: the penicillin type (produced by Penicillium spp.) and the cephalosporins (produced by Cephalosporium acremonium Streptomyces spp.). The penicillin-type fermentation is characterized by the production of a range of penicillins with nonpolar side chains, each produced by utilization of an appropriate precursor; 6-APA and isopenicillin N accumulate when the source of appropriate side chains is minimal. Similarly, cephalosporin-type fermentations produce penicillin N (having the C-aminoadipic acid side chain) as the sole penicillin. But they give rise to many cephalosporins and cephamycins.

Thus, experience with penicillin made possible the difficult isolation of the cephalosporin nucleus (7-aminocephalosporanic acid) and the subsequent development of numerous cephalosporin derivatives. The way was clear for the harnessing of knowledge of biosynthetic pathways in antibiotic-producing organisms, and for further discoveries of new semisynthetic cephalosporins with specific biological activities.

12

Impact of Penicillin on Science and Medicine

Penicillin was the first antibiotic developed to a stage allowing its controlled trial as a therapeutic agent for parenteral treatment of infections in humans. In 1940 it was, and it still is forty years later, the most active antimicrobial agent ever described.

No one questions the impact that penicillin has had on science and medicine. Yet it is sobering to note that penicillin might not exist today if 1980 to 1983 concepts of ethical practices and 1980 to 1983 rules and regulations governing studies of new agents in human beings had prevailed in 1940 to 1942. Early preparations of penicillin were impure and highly pigmented; they contained pyrogenic materials that would not be allowed in any pharmaceutical product today. Yet this crude material produced in people such dramatic therapeutic responses that the expenditure of massive amounts of human time and talent, precious supplies and monies, seemed justified.

Streptomycin, the second widely used antimicrobial drug, was discovered four years after the Oxford group of investigators had first reported the isolation of penicillin in a crude but usable form. Streptomycin was active against a different spectrum of infecting organisms, and it was less active on a weight basis, more toxic, and more prone to allow the emergence of drug-resistant organisms than was penicillin. Yet it quickly became an accepted chemotherapeutic drug for treatment of those infections caused by microorganisms susceptible to its action.[1] Knowledge of all that penicillin could do made it easy to believe that another antimicrobial drug could do likewise. Experience with penicillin had provided the base necessary for the development of streptomycin and, later, polymyxins,

PLATE 12.1.   Original solvent extractor for penicillin contrasted with large modern extractor, Pfizer, Inc.

neomycins, chloramphenicol, tetracyclines, and the many other antimicrobials that followed.

Early interest in penicillin derived from the medical profession's need for a means of treating infectious diseases. Unquestionably, World War II stimulated and speeded its development. But even without such a stimulus, penicillin—or another agent—would soon have been introduced for treatment of infection. It would have taken longer for it to gain acceptance; penicillin might not have been first but the need existed. Interest in bacterial antagonisms and knowledge of the phenomenon of antibiosis were pointing the way.

Dr. Vannevar Bush, director of World War II's Office of Scientific Research and Development, was asked by President Franklin D. Roosevelt in November 1944 to advise on how to organize a program for continuing the work that had been done in medicine and the related sciences during the war.[2] Roosevelt requested specifically an effective program for discovering and developing talent in American youth so that scientific research in the United States would continue to function on a level comparable to that of World War II. Bush based his report primarily

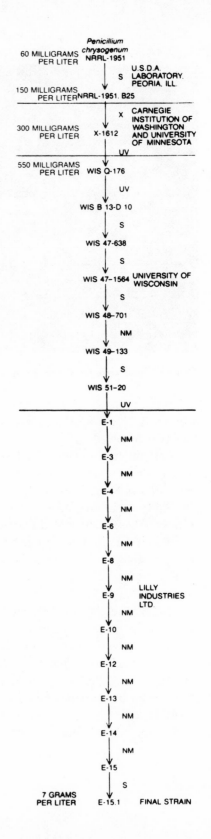

FIGURE 12.1.   Repeated mutations were necessary to create a strain of the mold Penicillium chrysogenum *that synthesized enough penicillin to form the basis of a commercial process. Radiation and a chemical agent were employed by four groups of investigators to induce mutations in the mold. ("S" stands for spontaneous mutation, "X" for X-radiation, "UV" for ultraviolet radiation, and "NM" for nitrogen mustard.) Selection of the superior strains ultimately gave rise to strain F.15.1, which yielded 55 times as much penicillin as laboratory strains. Simultaneous improvements in fermentation technique increased yields still further; yield figures in this chart reflect both kinds of increase. Classical genetic techniques such as these are still important in the antibiotics industry. Complexity of antibiotic synthesis in microorganisms makes it impractical to develop new strains by directly altering single genes. Current fermentation methods yield more than 20 grams per liter. (Reproduced with permission from Y. Aharonowitz and G. Cohen, "The Microbiological Production of Pharmaceuticals," Scientific American 245, no. 3 [September 1981]: 144. Copyright [1981] by the Scientific American, Inc. All rights reserved.)*

on the natural sciences, and the program he presented to President Harry Truman (who by then had succeeded Roosevelt) laid the foundation for much that transpired in the United States in subsequent years. The development of penicillin figured strongly in Bush's thinking and in his proposals; the establishment of the National Science Foundation was a direct outgrowth of this report.

Penicillin was developed at a time when investigators in industry, universities, and government were working closely together and the distinctions between applied and basic research were beginning to blur. Relations between universities and industry peaked as research on penicillin began to peak. Clinicians, biologists, and chemists at the university level needed the talent, the equipment, and the production know-how that industry could provide. They needed the impetus and dollars that government could provide as well as the scientific expertise available through government facilities and personnel. Thus, the development of penicillin became a prime example of what collaboration among government, industry, and the universities could achieve.[2]

Penicillin and Infectious Disease

The antimicrobial drugs available for clinical use in 1941 were quinine, quinacrine, the arsenicals, and the sulfonamides.[3,4] Only the sulfonamides were active against bacterial infections, and even these had a narrow range of utility because of their toxicity. By 1945, penicillin was in widespread use, and by 1960, seven classes of antimicrobial drugs had been identified: the beta-lactam antibiotics (including the penicillins and cephalosporins), aminoglycosides (including streptomycin, among other drugs), macrolides, ansamycines, polypeptide and depsipeptide antibiotics, and a few miscellaneous agents that did not fall into any of these categories. Together, these drugs brought about major changes in the incidence, mortality, and epidemiology of infectious diseases worldwide. By 1960, the world had seen a remarkable increase in the capacity of medicine to alter the course of many infectious diseases (see figures 12.2 and 12.3).[3]

In 1982 Dr. Walsh McDermott attempted to convey his recollections of infectious diseases in the "before and after" penicillin period and called attention to the fact that few physicians now in practice can recall these diseases as they were in the 1930s and before. According to McDermott—and this was certainly true—it became clear almost immediately after penicillin was available for clinical use that lives could be saved by the proper administration of the drug. But time was required to establish

235

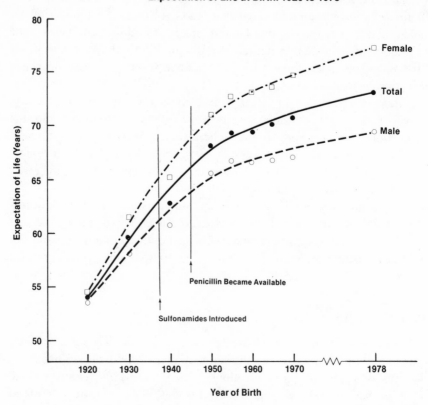

Expectation of Life at Birth: 1920 to 1978

FIGURE 12.2.   Life expectancy at birth, 1920–1978. (From: Statistical Abstract of the United States, *101st ed., 1980, Department of Commerce, Bureau of the Census, table 106, "Expectation of Life at Birth: 1920–1978.")*

changes in epidemiology and in those disease characteristics that influenced responses to chemotherapy:

The bringing of each disease under control was in itself a separate miracle. These were thrilling events. If traditionally the medical profession was slow to embrace the new, it soon lost all such shyness. Therapeutic triumphs coming rapidly one after another imprinted the physicians with almost too great a readiness to believe. Faced with the tale of some new remedy, an impulse to seek proof would arise from the science-based portion of their education, only to be met by the internal rejoinder, "why not believe?" And, among the most poignant and enduring recollections would be those of the distressing terminal illness of the last patient on "the old" treatment, as contrasted with the "recovery" of the first patient on the "new" treatment...[gradually there] were changes in the ways...doctors

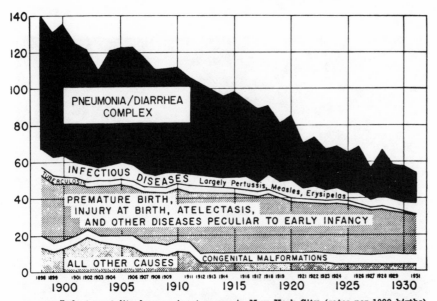

Infant mortality by prominent causes in New York City (rates per 1000 births)

Source: *Weekly Reports of the Department of Health, New York City,* Vol. XXI, no. 50, p. 396, December 17, 1932.

FIGURE 12.3. Infant mortality by prominent causes in New York City (rates per 1,000 births), 1900–1930. (From: W. McDermott, "Pharmaceuticals: Their Role in Developing Societies," Science 209, no. 4453 [July 11, 1980]): 240–45. Copyright 1980 by the American Association for the Advancement of Science.)

practiced medicine and the extension of...things possible.... The impact of [antimicrobial] technology, and it has been great, has been on the extent to which by providing a potential control for infection it has expanded what...[physicians] can do.[3]

The use of penicillin and the antimicrobial drugs that followed affected virtually all aspects of medicine. Not only did it decrease the incidence and severity of many infectious diseases, but, as McDermott reminded us,[3] it made possible cardiac surgery, organ transplantation, and the management of severe burns. It decreased mortality due to hemolytic streptococcal infections, such as puerperal sepsis, and it changed the frequency and nature of infectious diseases seen in the practice of pediatrics. It also decreased mortality due to illnesses such as pneumococcal pneumonia, otitis media, and bacterial meningitis (though pneumococci, for example, remain prevalent today and are a major cause of community-acquired bacterial pneumonia—the incidence of which

PLATE 12.2.   Prof. Walsh McDermott, M.D., D.Sc. (Hon.), who devoted a
lifetime to the study of antimicrobial drugs and their impact worldwide on society and
medicine. (Courtesy of Mrs. Walsh McDermott and the Robert Wood Johnson
Foundation. Photograph by Orren Jack Turner.)

approximates one-half million cases annually in the United States—and otitis media in infancy and early childhood—the incidence of which exceeds one million cases yearly).[4]

Pneumococci are among the organisms most susceptible to the action of penicillin. Yet they often persist for long periods of time following apparently adequate chemotherapy of infected hosts. Despite the remarkable antibacterial activity of penicillin, eradication of penicillin-susceptible organisms from infected persons occurs rarely, if ever. Microbial persistence, although difficult to detect, occurs frequently. The survival of drug-susceptible microorganisms within the host despite the presence of high concentrations of drug is a phenomenon that, once described, provided a basis for understanding the mechanisms by which late relapse may occur in humans.[5]

As long as an infectious agent, whether a pneumococcus or streptococcus, a tubercle bacillus or other living organism, exists within an individual, that individual is infected. One cannot prevent infection, nor, at least from the microbiologist's viewpoint, can one easily eradicate infection. One can hope only to prevent the development of disease in the infected host.

PLATE 12.3.   Modern analytical equipment used to analyze and control fermentation processes.

Microbial Resistance to the Action of Penicillin. The development of resistance of bacteria to commonly used antibiotics is one of the most important factors limiting the effectiveness of antimicrobial therapy today. Organisms resistant to the sulfonamides and to penicillin were described promptly after their first use in clinical medicine. Yet many organisms such as the pneumococci, hemolytic streptococci, and gonococci remained adequately susceptible to benzyl penicillin for many years. The introduction of antimicrobial drugs initially active against some other organisms, particularly the *Enterobacteriaceae*, however, brought with it other problems. The spread of R-factors (resistance factors) among the *Enterobacteriaceae* and of resistance plasmids into unrelated bacterial species was the cause of deep concern.[6]

Penicillin resistance, however, occurs differently. It relates to the capacity of an infecting organism to produce an enzyme, penicillinase (beta-lactamase), which destroys penicillin's antimicrobial effectiveness. This obviously limited penicillin's usefulness over the years. Recently introduced penicillinase-inhibiting adjuvants as a class, however, are now having a great impact on the treatment of infections caused by beta-lactamase-producing organisms.[7,8] In mixed infections, moreover, where penicillinase-producing nonpathogens may protect pathogenic organisms from penicillinase-susceptible (unstable) drugs, the use of an antibacterial drug in combination with an inhibitor of the enzyme that destroys it offers promise.[9]

The Changing Nature of Infectious Diseases. During the past twenty years (i.e. in the 1960s and 1970s), interest in infectious diseases waned to some extent. Surely this is true of those diseases caused by antibiotic-susceptible microorganisms. It is difficult to retain interest in illnesses that do not occur or are easily treated even though historically they may have been most important at one time. It is difficult to retain interest in diseases one does not see, does not hear about, or in new and pathogenic microorganisms isolates of which one cannot obtain for study. But infection has not vanished and the prevalence of infectious diseases as a class has not waned. In 1980, there were at least two hundred known infectious diseases that could neither be treated nor prevented.[10] Most were caused by viruses which mutate so easily that the development of specific and effective agents for their control has been difficult. Some are due to disease-producing agents that have long been known; some appear to be due to previously undescribed agents of disease.

Ironically, penicillin and the antibiotics that followed are responsible for the current lack of interest in infectious diseases. This indifference is a

PLATE 12.4.   Manufacture of penicillin is accomplished in a combination of biological and chemical steps; shown here are crystallizers, the site of one of the key processes in production. Manufacture begins in fermentation tanks with a capacity of as much as 100,000 liters. In the tanks, an industrial strain of the fungal mold Penicillium chrysogenum is grown in a rich liquid medium; a form of penicillin called penicillin G is a natural metabolite of the fungal cells. In this plant, operated by Pfizer, Inc., in Groton, Connecticut, as many as 15 fementation tanks are linked on a staggered production schedule to provide a continuous output of the antibiotic. When fermentation, which takes several days, is complete, penicillin G is separated from the spend mold cells and injected into the crystallizers, where butanol is added. The butanol is evaporated, carrying water with it and leaving behind a crystalline slurry of penicillin G of more than 99 percent purity. Subsequent chemical modifications yield other forms of penicillin. (Photographed by Ralph Morse, Rockaway, New Jersey. Reproduced with permission from: Y. Aharonowitz and G. Cohen, "The Microbiological Production of Pharmaceuticals," Scientific American 245, no. 3 [September 1981]: 140. Copyright 1981 by the Scientific American, Inc. All rights reserved.)

measure of the success of antibiotics in the treatment and control of infection. Yet, lest we relax too much, we must remember the remarkable capacity of microorganisms to persist, to mutate, and thus to thwart the efforts of the best of us.

Penicillin and Industrial Microbiology

The pharmaceutical industry initially focused primarily on the production of penicillin by fermentation. Later, as other antibiotics were discovered, they too were produced mostly by fermentation. Unquestionably, today, antibiotics are among the most important pharmaceutical products manufactured in part or in full by the growth of living microorganisms.

Antibiotics are secondary metabolites, the raw materials for production of which are primary metabolites that result from the growth of molds (or occasionally other organisms). Initially produced in small amounts only, some antibiotics detected were active in vivo, some not; some appear to have other uses. Increasing yields of the usable products of fermentation became the concern to those involved in penicillin production. But with time, it was shown that the proper combination of nutritional substances in culture media and genetic manipulations may lead at times to remarkable primary metabolite overproduction. The introduction of penicillin and, later, other antibiotics brought about an explosive increase in the number of products commercially produced by microbial growth and usable in medicine, nutrition, industry, and research.[13–15]

The mechanisms involved in the production of antibiotics and the factors limiting their biosynthesis are of major interest. The same genetic mechanisms control production of primary and secondary metabolites, and mutation and screening for high-yield cultures (ones that produce high yields of desired compounds) have been most successful. One-hundred- to one-thousand-fold increases in antibiotic production have been achieved, and the yields of penicillin at the end of a fermentation cycle may at times approximate 40,000 or more units per milliliter. Methods for production of this large amount of penicillin have evolved gradually over many years, and because of the diffusion of technology, they are fairly uniform throughout the industry.[11–18]

Enzyme biosynthesis, the bioconversion of antibiotics by one or more enzymes to structurally related antimicrobials, and the production of single-cell protein for use as animal feed have all evolved from the new technologies that emanated from needs of the penicillin industry. Because

PLATE 12.5.   Inside of large modern fermentor, showing a man at the bottom of the tank to give an idea of size. Cooling coil consists of more than one mile of stainless steel pipe.

deoxyribonucleic acid (DNA) plays a fundamental role in all living processes, genetic manipulation affects the development of any industry employing microorganisms or their products.[12,17] Six genera of filamentous fungi have given rise to hundreds of recognized and distinct antibiotics. The introduction of recombinant DNA technology, moreover, makes feasible the synthesis of many other substances of commercial importance. And the patentability of microbial strains makes their development worthwhile.[19-20]

Antimicrobial Agents: An Interdisciplinary Science

Thus, from penicillin there has evolved a new and interdisciplinary science relating to products of microbial fermentation. Perhaps first recognized during World War II when persons of all disciplines were enlisted to speed production, later recognized by those concerned with the medical use of antibiotics, it is now clear that all who are concerned with antibiotics and with industrial or applied microbiology must approach problems from an interdisciplinary point of view. Essential is the joint participation of biologists, including chemists, clinicians, pharmacologists, and engineers. The role of genetics and plasmids moreover holds the spotlight today. No longer can one rely on chance observations or chance events. We have come a long way since Alexander Fleming first observed the lytic action of penicillin in 1928. Scientific knowledge and reason now provide the basis for each scientific advance. Moreover, a major concern now is the way in which the scientific and economic progress that emanated from Fleming's discovery affects the quality of life. As F. Michael Lee, deputy director of Health Economics in England, commented in 1965:

In looking at science in the economy, we must replace our quantitative preconceptions...by more complex qualitative notions of a changing, not merely a rising standard of living. In economic growth we are not concerned merely with greater efficiency leading to the accumulation of material wealth, but also with how scientific and economic progress affects the quality of life. From any point of view, the most significant change in pharmaceutical production in recent decades is not that penicillin prices are now only [a fraction of what they were] or that [another antibiotic] could be obtained more cheaply; it is that the products now exist.[22]

In 1984 this seems even more true. As new antimicrobial agents have come to the fore during the past twenty years, the quality of life has continued to change for the better. What still stands out as most important is that these agents continue to exist.

Epilogue

Many books have been written on the history of penicillin. Each is different, for each reflects its author's personal reasons for telling the story.

The present book is not complete nor will any book on penicillin ever be—the subject is too vast. I have tried, however, to touch on as many aspects as possible. I have written the book entirely myself. If there are errors, omissions, or misrepresentations, they are mine and I apologize to those concerned. I have tried, however, to present in a factual and unbiased way what my review of the literature and my personal experiences lead me to believe actually transpired. Inevitably, the book reflects my personal interests; thus, the reader's attention is directed primarily to the biological aspects of penicillin and its development. Perhaps this can be justified on the basis that penicillin is a biological product—produced by living organisms and important only because of its remarkable biological effects. Perhaps it may be justified too by the fact that its development as a chemotherapeutic agent was accomplished almost entirely through advancements in biological technology.

As one reviews what transpired between 1941 and 1945, one cannot help being amazed at the size of the project, the number of persons involved, the diverse nature and large number of organizations that contributed to its success, and the remarkable organizational ability of Vannevar Bush, A. N. Richards, and those who worked with them to create and administer what was then—and probably still is—one of the greatest research ventures ever carried to fruition. It is rivaled only by the Manhattan Project, also conducted during World War II.

I have attempted to document all that I have included in this book. Unfortunately, however, as I mentioned in the preface, documentation destroys the flavor of the story. Those who worked with penicillin during its early developmental period will all agree, I feel sure, that it was the most exciting and rewarding period of their professional lives. Yet I doubt that the excitement, pleasure, even pride that we felt—as each step of the process was solved and each report of lives saved by the drug reached us—is readily apparent in this account.

Documentation also eliminates the spontaneity that was so much a part of the penicillin developmental program. Techniques were not so refined then as now. Improvising was essential, a fact rarely mentioned in historical reviews of penicillin. The resourcefulness and ingenuity of those involved often made the difference between success and failure. That resourcefulness, moreover, often led to great fun and was an important aspect of the program.

Consider a still for evaporation of solvents on the fire escape of a high-rise building in the heart of New York City; an "incubator room" under the seats of an amphitheater; rabbits with experimental syphilis "home for an Easter weekend" so that their penicillin injection schedules would not be interrupted; the reaction of top executives to attempts to recover penicillin from fermentation liquors by freezing out water ("an expensive way to make ice"). Humorous episodes of this sort made the struggle to produce penicillin pleasurable.

Consider, too, Penny Mehler—a girl twelve to fourteen years of age on the pediatric service of the Presbyterian Hospital in New York City with subacute bacterial endocarditis—who was one of Dawson's and Hunter's first cures, or Miriam Laskowitz, another of their first patients treated for the same illness, who commented to me many years later that she had had surgery for all sorts of reasons, but "my heart is fine." And consider also Otto Morowitz, who in 1944 was reported in the *Journal of the American Medical Association* as a treatment failure. It was so obvious that he would not survive that he was reported as having died, yet in 1978 he was remarkably well, having just the previous year survived a double valve replacement. It was episodes of this sort that made the struggle to produce penicillin so very worthwhile.

Many have claimed that the development of penicillin engendered rivalries and hatreds among those who worked on it in the early 1940s. "Disputes" between Alexander Fleming and Howard Florey have been cited repeatedly—indeed magnified out of proportion—as have differences between Florey and Ernst Chain. I have said nothing of these differences in this book. Such jealousies and differences of opinion, except as they relate to scientific facts or observations, are not a part of the penicillin story. Even in the early 1940s, everyone knew that penicillin was of great importance. The results obtained in each phase of its development were astonishing, and everyone quite naturally wished to receive a full share of the credit for what was accomplished.

Today it is clear that, just as the development of penicillin engendered some jealousies and rivalries among investigators, so too it led to many close friendships among scientists in the United States, England, and elsewhere. It led many to make the study of penicillin and the antibiotics that came later a lifetime venture. It became the base on which the science of antimicrobial action and chemotherapy rests today.

Recently (in "New Light on the History of Penicillin," *Medical History*, no. 26 [1982]: 1–24), Dr. Ronald Hare raised a question that has been asked many times: why didn't Fleming himself establish the therapeutic efficacy of penicillin? In this book, I have expressed the opinion that the scientific community was not ready for penicillin in 1928–29. It was not ready to perform or even permit the experimental procedures necessary to establish its therapeutic efficacy.

It should be remembered that Fleming was born in 1881. He was forty-seven years old when he discovered penicillin and near sixty when Florey first called attention to its potential. The anticipated average life span at the time of birth for those born in 1920 was only about fifty-four years; for those born in 1881 it was far less. Fleming, then, had almost lived his assumed full life span by the time he discovered penicillin.

From 1945 on, from the time when he along with Sir Howard Florey and Dr. Ernst Chain received the Nobel Prize for the discovery and development of penicillin, Fleming was feted by royalty, the medical profession, the U.S. pharmaceutical industry, the press, and the lay public alike. As he once said to me, one can rebel against such publicity and such favors or one can gracefully accept and enjoy them. At the age of sixty-four in 1945, he chose the second path. Fleming accepted the emotional reactions of the lay press and the public (who by then knew the importance of penicillin) with good humor and quiet amusement. Howard Florey, on the other hand, who was considerably younger, chose to reject all but the most important accolades—and to get on with the work at hand.

One may ask, who among us has ever performed the complete experiment, leaving nothing for others to do? Each of us starts with a simple observation, perhaps even a "discovery," and all else follows. One person after another builds on the original observation, ultimately bringing scientific advancement and success. Each does what is possible in his time. Had Fleming not made the observation and preserved the culture, had he not demonstrated the presence of penicillin in his culture fluids and recorded its properties, there would have been no starting point for Chain and Florey's studies ten years later. And had these two not recognized penicillin's potential and persisted until it was available for clinical use, one may wonder what the aftermath of World War II would have been.

One cannot doubt that discovery is always an expression of the intellectual, social, and economic pressures in the environment in which it is born.

René Dubos, Louis Pasteur—Freelance of Science (1950)

Appendix

Twenty-one antibiotics facilities were authorized by the U.S. government in 1943 to 1945, seventeen for penicillin production, one for penicillin research, and four for production of penicillin and other products. The total cost was $30,218,367. Of this, $7,577,697 were public funds; $22,640,670 came from private funds (63.9 percent of which were authorized for rapid tax amortization). Of the $7,577,679 expended by the government, $3,352,724 (44.2 percent) were recovered on later sale of six of the World War II penicillin plants. By 1956, only twelve of the firms that were involved in penicillin production in 1943 to 1945 were still manufacturing the drug. Antibiotics research contracts were concluded with fifty-eight institutions during the 1943 to 1945 period, at a cost of $2,770,305.

TABLE A.1   Output of leading antibiotics and percent to total output: 1948 and 1956

Antibiotic	1948		1956	
	Number of Pounds[1]	Percent of Total	Number of Pounds[1]	Percent of Total
Penicillins	155,873	64.9	1,059,704	34.4
Streptomycin	80,737	33.6	148,999	4.8
Dihydrostreptomycin	2,989	1.2	492,173	16.0
Chlortetracycline	661	.3	560,663	18.2
Oxytetracycline			324,614	10.5
Tetracycline			220,074	7.1
Chloramphenicol	46	([2])	85,408	2.8
Erythromycin			70,913	2.3
All others	26	([2])	118,825	3.9
Total	240,332	100.0	3,081,373	100.0

[1]Penicillin has been measured in international units of potency (Louis S. Goodman and Alfred Gilman, *The Pharmacological Basis of Therapeutics* [New York: Macmillan Co., 1956], p. 1239). In this study, conversion to pounds is done on the basis of published potency ratios. These ratios vary from one penicillin salt to another and do not always conform to the ratios used by the Tariff Commission in its annual reports.

[2]Less than 0.1 percent.

Source: FTC data requests, 1956 and 1957.

From: Federal Trade Commission, *Economic Report on Antibiotics Manufacture* (Washington: U.S. Government Printing Office, June 1958), p. 67.

TABLE A.2 Public funds expended and amounts recovered by sale of six World War II penicillin plants

Purchases and location of plants	Plancor No.[1]	Government expenditure[2]	Reported sales price Amount	Percent of cost to Government
Abbott Laboratories, North Chicago, Ill.	2271	$197,497	$136,258	69.0
American Cyanamid Co., Pearl River, N.Y.	1937	784,528	91,366	11.6
Ben Venue Laboratories, Bedford, Ohio	1942	335,672	150,000	44.7
Bristol Laboratories, Inc., East Syracuse, N.Y.	1935	2,410,000	1,100,100	45.6
Cutter Laboratories, Berkeley, Calif.	1926	850,000	175,000	20.6
Heyden Chemical Corp.,[3] Princeton, N.J.	1951	3,000,000	1,700,000	56.7
Total, 6 plants		7,577,697	3,352,724	44.2

[1]Numbers assigned by the Reconstruction Finance Corporation to plants in the program of the Defense Plants Corporation.
[2]From pp. 54–55, Federal Trade Commission, Economic Report on Antibiotics Manufacture, June 1958.
[3]This plant was used for the production of formaldehyde and hexamine in addition to penicillin.

Sources: War Assets Administration, Office of Real Property Disposal, Sales and Leases of Industrial, Transportation, and Maritime Property as of June 30, 1947, Washington, D.C.
Records of the Reconstruction Finance Corporation, NA, RG 234, National Archives, Washington, D.C.
General Services Administration, Public Buildings Service, Surplus Industrial Real Property and Related Personal Property, Sales and Transfers as of September 30, 1952, Washington, D.C.
Annual reports of antibiotics manufacturers.

From: Federal Trade Commission, Economic Report on Antibiotics Manufacture, U.S. Government Printing Office, June 1958, p. 52

TABLE A.3 Antibiotics facilities authorized during 1943–1945

Name of applicant—location of facilities	Type of facility		Total cost	Public funds	Private funds	Percentage of private funds authorized for rapid amortization	Amount approved for rapid tax amortization	Facilities completed (year)
	Production	Other						
Abbott Laboratories, North Chicago, Ill.	Penicillin		$670,704	$197,497	$473,207	100.0	$473,207	1944–45
American Cyanamid Co. (Lederle), Pearl River, N.Y.	-do-		3,170,888	784,528	2,386,360	87.0	2,075,230	1944–45
American Home Products Corp. (Wyeth):[1]								
Kimberton, Pa.	-do-		470,082		470,082	99.4	467,216 ⎫	1943–45
West Chester, Pa.	-do-		506,322		506,322	85.3	431,962 ⎬	
Philadelphia, Pa.	-do-	Research penicillin	280,000		280,000			
Ben Venue Laboratories, Bedford, Ohio	-do-		335,672	335,672				1944
Bristol Laboratories, Inc., East Syracuse, N.Y.	-do-		2,410,000	2,410,000				1944
Cherokee Biological Laboratories,[2] Syracuse, N.Y.	-do-		210,000		210,000			n.a.
Commercial Solvents Corp., Terre Haute, Ind.	-do-		1,869,000		1,869,000	97.2	1,817,000	1944
Cutter Laboratories, Berkeley, Calif.	-do-		850,000	850,000				1944

Company	Products					
Emerson, Sam W., and Hattie Dettleback.[2] Cleveland, Ohio			36,000	36,000			n.a.
Heyden Chemical Corp., Princeton, N.J.	-do-	3,000,000	3,000,000			1943
Hoffmann-LaRoche, Nutley, N.J.	-do-					n.a.
	Penicillin, ascorbic acid, thiamine chloride, riboflavin, synthetic vitamins	1,935,000	1,935,000	66.2	1,281,000	
Eli Lilly & Co., Indianapolis, Ind.	Penicillin	710,880	710,880	77.1	548,079	1944–45
Merck & Co., Inc., Rahway, N.J.	Penicillin, thiamine hydrochloride, synthetic papaverine, sulfathiazole, sodamide, ascorbic acid, acetylaminobenzene, sulfomyl chloride, niacinamide, riboflavin, vitamin B-6, sulfapyridine, atabrine, codeine, dichlorodiphenyl, trichlorethane, flour-enriched tablets	5,749,000	5,749,000	4.8	276,000	n.a.
Merrell, William & Co. (Vick),[2] Hamilton, Ohio	Pilot plant: streptomycin	65,000	65,000	100.0	65,000	n.a.
Parke, Davis & Co., Detroit, Mich.	Penicillin, epidermic typhus vaccine, gelatine capsules, influenza virus vaccine	789,000	789,000	60.8	480,000	n.a.

(*Continues*)

TABLE A.3 (Continued)

Name of applicant—location of facilities	Type of facility Production	Type of facility Other	Total cost	Public funds	Private funds	Percentage of private funds authorized for rapid amortization	Amount approved for rapid tax amortization	Facilities completed (year)
Chas. Pfizer & Co., Inc., Brooklyn, N.Y.	Penicillin		3,396,902		3,396,902	92.8	3,151,396	1944–45
Schenley Laboratories, Inc., Lawrenceburg, Ind.	-do-		1,175,174		1,175,174	100.0	1,175,174	1943–45
E. R. Squibb & Sons,[3] New Brunswick, N.J.	-do-		2,107,615		2,107,615	88.9	1,874,128	1945
Sterling Drug, Inc., Rensselaer, N.Y.	Penicillin, atabrine, and sulfathiazole		157,000		157,000	51.0	80,000	n.a.
Upjohn Co., Kalamazoo, Mich.	Penicillin		324,128		324,128	83.2	269,590	1944–45
Total			30,218,367	7,577,697	22,640,670	63.9	14,464,982	—

"n.a." indicates that data are not available.

[1]Includes Reichel Laboratories purchased by American Home Products, December 1, 1942.

[2]See source note, ch. 1, to Civilian Production Administration list of companies authorized to receive priority assistance.

[3]Merged with Olin Mathieson in 1952.

Sources: FTC data request 1957, schedule VIII, and Civilian Production Administration, Industrial Statistics Division, War Industrial Facilities Authorized, July 1940–August 1945, Washington, July 30, 1946.

Reproduced with permission from: Federal Trade Commission, Economic Report on Antibiotics Manufacture, U.S. Government Printing Office, June 1958 report, pp. 54–55.

TABLE A.4 Antibiotics research contracts concluded by the Office of Scientific Research and Development (OSRD) during World War II

Contractor	Contract No.[1]	Subject of research (short title)	Cost to Government
Bradley Polytechnic	100	Culture Methods and Assay of Penicillin	$27,121.85
Princeton University	119	Method for Quick Titration of Potency of Penicillin	3,000.00
St. Louis University	155	Discovery of Products Superior to Penicillin and Pyocyanase	2,001.00
Massachusetts Memorial Hospitals	275	Collection and Coordination of Information Concerning Penicillin	1,834,444.70
University of Buffalo	359	Penicillin Therapy of War Wounds	3,408.09
E. R. Squibb & Sons	389	Chemical Structure and Synthesis of Penicillin	(?)
Chas. Pfizer & Co., Inc.	390	-do-	(?)
Merck & Co., Inc.	391	-do-	(?)
Johns Hopkins University	393	Effect of Penicillin in Syphilis	91,434.50
Abbott Laboratories	396	Chemical Structure and Synthesis of Penicillin	(?)
Eli Lilly & Co.	397	-do-	(?)
Parke, Davis & Co.	398	-do-	(?)
The Upjohn Co.	399	-do-	(?)
University of Pennsylvania	403	Effect of Penicillin on Syphilis	100,898.00
New York University	404	Treatment of Early Syphilis with Penicillin	61,057.00
University of Michigan	408	Chemical Structure and Synthesis of Penicillin	68,040.13
U.S. Department of Agriculture	410	Chemical Structure of Penicillin	46,483.00[3]
Cornell University	411	Chemical Structure and Synthesis of Penicillin	162,112.53
University of Illinois	426	Value of Penicillin in Treatment of Compound Fractures	6,125.00
Tufts College	427	Treatment of Compound Fractures with Prophylactic Penicillin Therapy	17,223.90

(Continues)

TABLE A.4 (Continued)

Contractor	Contract No.[1]	Subject of research (short title)	Cost to Government
Winthrop Chemical Co. } Heyden Chemical Corp. }	428	Chemical Structure and Synthesis of Penicillin	([2])
University of California	431	Use of Penicillin in Infections, Especially Craniocerebral Infection	16,792.12
Cornell University	435	Effectiveness of Penicillin Therapy in Syphilis	14,407.46
University of Illinois	439	Structure of Penicillin and Related Compounds	8,689.25
University of Michigan	442	Infrared Pictures of Penicillin Crystals	39,892.16
Shell Development Co. } Cutter Laboratories }	445	Chemical Structure and Synthesis of Penicillin	([2])
Duke University	446	Penicillin Treatment for Syphilis	5,725.15
Hospital for Joint Diseases	457	Use of Penicillin in the Therapy of Hematogenous Osteomyelitis	6,500.00
Washington University	460	Effect of Penicillin on Character and Severity of Bacterial Infections	3,210.08
Tulane University	461	Treatment of Congenital Syphilis with Penicillin	5,250.00
Vanderbilt University	462	Clinical Value of Penicillin in Various Infectious Conditions	6,092.00
Food and Drug Administration	465	Improvement in Assay Procedures Used in Control of Penicillin	([4])
Mount Sinai Hospital	479	Penicillin Therapy of Subacute Bacterial Endocarditis	3,525.00
Wills Hospital, Philadelphia	480	Penicillin Treatment of Ocular Syphilis	1,462.00
University of Virginia	495	Penicillin in the Treatment of Syphilis	4,008.00
University of Texas	496	Use of Penicillin in Treatment of Syphilis, Acquired and Congenital	10,863.00
Western Reserve University	497	Penicillin Therapy of Syphilis	7,842.03
University of Chicago	501	Effects of Penicillin Therapy on Central Nervous System	3,511.75
Vanderbilt University	505	Effectiveness of Penicillin in Treatment of Syphilis	1,612.94
Chicago Board of Health	510	Penicillin in Venereal Disease (Syphilis)	49,444.32
Wayne University	511	Penicillin Therapy of Syphilis	8,352.93

Institution	No.	Project	Amount
Stanford University	514	Treatment of Early Syphilis with Penicillin	6,249.94
University of Michigan	515	Penicillin Therapy for Early and Late Syphilis	4,325.00
Washington University	519	Treatment of Early Syphilis with Penicillin	5,697.25
University of California	520	Penicillin Treatment of Subacute Bacterial Endocarditis	2,252.89
Johns Hopkins University	526	Efficacy of Derivatives from Penicillin in Treatment of Experimental Syphilis	4,465.00
-do-	527	Biostatistical Analysis of Nationwide Results of Penicillin Therapy in Syphilis	33,737.00
Harvard University	540	Chemical Structure and Synthesis of Penicillin	13,546.83
Cornell University	542	Compilation of Chemical Index in Connection with Patient Questions	11,520.00
Merck Institute	544	Properties of Streptomycin	1.00
Washington University	551	Pharmacologic and Toxicologic Study of Streptomycin	1,333.95
St. Louis University	553	Discovery and Development of Products Superior to Penicillin and Pyocyanase	1.00
New Britain General Hospital	558	Penicillin and Other Antibiotic Agents in Experimental Syphilis	9,765.23
Stanford University	561	Penicillin Biosynthesis and Inactivation in Culture	17,704.00
Massachusetts Memorial Hospitals	569	Prepare Material for a Monograph on Penicillin	9,176.48
National Academy of Sciences	571	Prepare Monograph on Work Done Under U.S. and U.K. Sponsorship on Penicillin	30,000.00
Total			$2,770,305.46

[1] OEM cmr series.

[2] No cost to the government.

[3] This sum represents a transfer between government agencies.

[4] Not available.

Sources: Advances in Military Medicines, vol. 11, E. C. Andrus et al., eds., Little, Brown & Co., Boston, 1948, pp. 831 ff.; Records of OSRD, NA, RG 227, National Archives, Washington, D.C.

Reproduced from: Federal Trade Commission, Economic Report on Antibiotics Manufacture, U.S. Government Printing Office, June 1958, pp. 48–49.

TABLE A.5   1956 status of World War II authorized manufacturers of antibiotics[1]

Manufacturing under same ownership	Manufacturing under new name or ownership	Manufacturing not undertaken or discontinued
Abbott Laboratories	Heyden Chemical Co. (Antibiotics Division) (now American Cyanamid)	American Home Products (Reichel)
American Cyanamid (Lederle)		Ben Venue Laboratories
American Home Products (Wyeth)	E. R. Squibb & Sons (now Olin Mathieson)	Cherokee Biological Laboratories
Commercial Solvents		Cutter Laboratories
Eli Lilly & Co.		Emerson & Dettleback
Parke, Davis		Hoffmann-LaRoche
Chas. Pfizer		Merrell, Wm., & Co. (now Vick Chemical Co.)
The Upjohn Co.		Schenley Distillers
Cheplin Biological Laboratories (Bristol)[2]		Sterling Drug Co.
Merck & Co.		

[1]The table lists nine war-authorized producers that discontinued manufacture. Reichel production facilities were combined with American Home Products' Wyeth operations; Ben Venue Laboratories and Sterling Drug Co. ceased production at the close of World War II; available data do not indicate that Cherokee Biological Laboratories, Wm. Merrell & Co., and Emerson & Dettleback ever produced on a laboratory or commercial scale; Cutter Laboratories last manufactured in 1953 (when production was only 6,385 pounds); Hoffmann-LaRoche last manufactured in 1949 (when total output was only 89 pounds); and Schenley last produced in 1952 (when total production was 25,426 pounds). J. T. Baker Chemical Co. (consolidated by Vick Chemical Co. after 1941) and Monsanto, not shown in table, entered the antibiotics industry after World War II and dropped out in 1953 and 1954, respectively. Both were relatively minor producers.

[2]Purchase by Bristol occurred during World War II.

Sources: Civilian Production Administration, Industrial Statistics Division, War Industrial Facilities authorized July 1940 through August 1945, Washington, July 30, 1946. FTC data request, 1957.

Reproduced from: Federal Trade Commission, Economic Report on Antibiotics Manufacture, U.S. Government Printing Office, June 1958, p. 83.

Notes

Chapter 1

1   A. Fleming, "On the Antibacterial Action of Cultures of a Penicillium, with Special Reference to Their Use in the Isolation of *H. influenzae*," *Brit. J. Exp. Pathol.* 10, no. 3 (1929): 226–36.

2   A. Fleming, "On a Remarkable Bacteriolytic Element Found in Tissues and Secretions," *Proc. Roy. S.*, ser. B., 93 (1922): 306–17.

3   E. A. Duchesne, *Contribution a l'étude de la concurrence vitale chez les micro-organismes: Antagonisme entre les moisissures et les microbes*. Dissertation, 1896. Army Medical Academy in Lyon, France. Alexandre Rey, Imprimeur de la Faculté de Mêdicine, −4, Rue Gentil, 4− (December 1897). His notes on the experiment read:

"1. Les moisissures (mucédinées) ne se développent pas, ou disparaissent, tout au moins, très hâtivement dans l'eau, sous un certain volume, et cela pour les principales raisons suivantes: a) l'exageration même de l'humidité; b) le mouvement de la masse liquide; c) enfin et surtout le résultat de la concurrence vitale.

"2. Il existe, en effet, un antagonisme très marqué et incontestable entre les moisissures et les bactéries qui ont été simultanément semées dans l'eau ou dans un liquide nutritif quelconque, et cet antagonisme tourne le plus souvent au profit des bactéries en ce qui concerne, tout tout au moins, les processus de vitalité et de végétalité.

"3. Si les microbes l'emportent ainsi presque constamment sur les moisissures, dans la lutte pour la vie, c'est par suite d'une plus grande résistance vitale et surtout d'une pullulation infiniment plus rapide due, elle-même, au phénomène de la bipartition ou scissiparité. Mais il ne semble pas que les toxines microbiennes soient appelées à jouer un rôle actif dans cette lutte et dans ses résultats.

"4. Les moisissures, cependant, peuvent parfois voir cette lutte tourner à leur profit lorsque le milieu de culture leur est, par sa réaction, plus nettement favorable qu'aux bactéries, qu'elles ne s'y trouvent pas absolument submergées et qu'elles sont enfin, initialement, en proportion vraiment très prépondérante.

"5. Il semble, d'autre part, résulter de quelques-unes de nos expériences, malheureusement trop peu nombreuses et qu'il importera de répéter à nouveau et de contrôler, que certaines moisissures (*Penicillium glaucum*), inoculées à un animal en même temps que des cultures très virulentes de quelques microbes pathogènes (*B. coli* et *B. typhosus* d'Eberth), sont capables d'atténuer dans de très notables proportions la virulence de ces cultures bactériennes.

"6. On peut donce esperer qu'en poursuivant l'étude des faits de concurrence biologique entre moisissures et microbes, étude seulement ébauchée par nous et à laquelle nous n'avons d'autre prétention que d'avoir apporté ici une très modeste contribution, on arrivera, peut-être, a la découverte d'autres faits directement utiles et applicables à l'hygiène prophylactique et à la thérapeutique."

4   R. A. Kyle and M. A. Shampo, "Ernest Augustin Clement Duchesne," *J. Amer. Med. Assoc.* 240, no. 9 (1978): 847.

5   A. Gratia, Letter to Dr. Robert Fulton, dated January 28, 1946, Fulton Collection (Penicillin) Box 65, Sterling Memorial Library, Manuscripts and Archives, Yale University.

6   Robert Koch, "Zur Untersuchung von Pathogenen Organismen," *Mittheilungen aus dem Kaiserlichen Gesundheitsamte* 1 (1881): 1–48.

7   Louis Pasteur, "Sur les virus-vaccins du cholera des poules et du charbon," *Comptes rendus des travaus du Congres International des Directeurs des Stations Agronomiques, session de Versailles* (June 1881): 151–62.

8   Robert Koch, "Die aetiologie der tuberkulose," *Mittheilungen aus dem Kaiserlichen Gesundheitsamte* 2 (1884): 1–88.

9   A. Fleming, "Antiseptics (The Lister Memorial Lecture)," *Chemistry and Industry*, no. 3 (1945): 18–23.

10 Joseph Lister, "On the Antiseptic Principle in the Practice of Surgery," *Brit. Med. J.* 2 (1867): 246.

11 Sir John C. G. Ledingham, "The Development of Medical Studies in Britain: V. Bacteriology," *Brit. Med. Bull.* 2, no. 12 (1944): 261–65.

12 Barnett Cohen, *Chronicles of the Society of American Bacteriologists, 1899–1950* (Baltimore: Williams & Wilkins Co. and Waverly Press, May 1950).

13 Sir Graham S. Wilson and Sir A. Ashley Miles, *Topley & Wilson's Principles of Bacteriology and Immunity*, 5th ed. (Baltimore: Williams & Wilkins Co., 1964), 1: 1–15.

14 A. Maurois, *The Life of Alexander Fleming* (New York: E. P. Dutton & Co., 1959).

15 L. Colebrook, "Alexander Fleming. 1881–1955," *Biographical Memoirs of Fellows of the Royal Society* 2 (1956): 117–27.

16 Ronald Hare, *The Birth of Penicillin* (London: George Allen & Unwin, 1970).

17 A. Fleming, "The Action of Chemical and Physiological Antiseptics in a Septic Wound," *Brit. J. Surg.* 7 (1919): 99–.

18 A. Fleming, "Penicillin—Its Discovery, Development, and Uses in the Field of Medicine and Surgery (The Harben Lectures, 1944)," *J. Roy. Inst. of Publ. Hlth. and Hygiene* (March–April 1945): 1–36.

19 A. Fleming, "A Comparison of the Activities of Antiseptics on Bacteria and on Leucocytes," *Proc. Roy. Soc.*, ser. B., 96 (1924): 171–80.

20 L. Colebrook, *Almroth Wright—Provocative Doctor and Thinker* (London: William Heinemann—Medical Books, Ltd., 1954). Fleming's method of evaluating the activity of antiseptics consisted of incubating blood that had been infected with small numbers of staphylococci with various concentrations of the antiseptic under study. Specifically, culture cells were made from two microscopic slides separated by means of five narrow strips of vaselined paper, arranged at intervals transversely to the long axis of the slides. By means of these strips of paper, the space between the two slides was divided into four very thin compartments or cells, open at each end. Each compartment could contain 50 c.mm. (i.e., 0.05 ml) or more of blood. If staphylococci were added to human blood in such quantity that there were about 100 organisms in each 50 c.mm. of blood, and if this mixture were then placed in the slide cells in 50 c.mm. (i.e., 0.05 ml) quantities, only a few (perhaps 2 or 3) organisms would be apparent after incubation. If, however, the bactericidal action of the leucocytes were destroyed before incubation, all the implanted staphylococci would grow. Thus, the effect of solutions of chemicals on the bactericidal power of leucocytes could be easily measured. This procedure allowed one to determine easily if the antileucocytic action of an antiseptic was greater (or less) than its antibacterial action.

21 A. Fleming, The Fleming Papers, British Library (London), Add. Mss. No. 56122, for the years 1940–45. Comment made at penicillin producers' dinner, June 25, 1945.

22 C. Thom, "Mycology Presents Penicillin," *Mycologia* 37, no. 4 (August 1954): 460–75. Retrospectively it is difficult to understand how Fleming, his mycologist (Mr. LaTouche), and his many associates at St. Mary's Hospital (who presumably saw the original plate with its contaminating mold) could all have mistaken the intense gray-green color of *Penicillium notatum* colonies with the reddish color of *Penicillium rubrum*. Sporulation may have been minimal on the original plate, but surely not on all subsequent subcultures.

23 A. Fleming, The Fleming Papers, British Library (London), 789E. Add. Mss. 56156 through 56166, for the years 1924–32. Some records are filed also at the Dick Institute in Kilmarnock, Scotland. See *n*33.

24 Fleming Archives, British Library (London), Collection 56209, Vol. CIV, Fleming 1928 (z.3.c.1.). Observations of colonies remaining on plates made by G. L. Hobby, 1979.

25 A. Fleming, The Fleming Papers, British Library (London), Add. Mss. 56156 through 56166, for the years 1924–32 (Vol. LVII, pp. 13–14, 1930).

26 Federal Trade Commission, *Economic Report on Antibiotics Manufacture*

(Washington, D.C.: U.S. Government Printing Office, June 1958), p. 304.

27 N. G. Heatley, Letter to G. Hobby, 1981. "Second edition, CMI catalogue (Commonwealth Mycological Institute, Ferry Lane, Kew)—*P. notatum West-ling.* 15,378. Contaminant of a culture, London, 1928. Sir Alexander Fleming's historical isolation. Fleming. Thom (144.5112.1) as *P. rubrum.* NRRL (824) G. Smith 1949 (=ACTC 9478).

"Seventh edition, CMI catalogue, 1975: gives much the same account of these cultures but adds, for No. 15,378: NCTC 4222, and omits 'Sir Alexander Fleming's historical isolation.'

"According to the Curator (Mr. Hill) of the NCTC, now at the Central Public Health Laboratories, Colindale, in North London: periodical lists of recent accessions to the NCTC, published in the *Trans. Brit. Mycol. Soc.* 19 (1935): 315–31 by St. John Brooks...Fleming's *P. notatum* was deposited on or before that date as NCTC 4222, but by whom was not stated.

"In the second edition of the CMI catalogue (see above), the name 'Fleming' at the end of the first line quoted [,i.e., following] '15,378, contaminant of a culture, London, 1928. Sir Alexander Fleming's historical isolation.' suggests that the culture was deposited by Fleming."

28 Letter from Dr. Chester S. Keefer to Dr. John F. Fulton, professor of history of medicine, Yale School of Medicine, dated July 12, 1944: "There is one statement in your letter which conveys an erroneous impression; namely, that Fleming used penicillin merely as a technical advantage in the culturing of soil organisms. Fleming was not interested in soil organisms but used penicillin to isolate gram negative organisms from mixed cultures. However, it is plain to anyone who reads the original paper [by Fleming] that he went far beyond that in his studies. He did everything except get a sufficient quantity of penicillin to use it intravenously. The original paper is a remarkable document, and I hope your historian who gives the factual account will read it carefully." Repro-duced from the original in the Yale University Library by permission of Mrs. John J. Fulton.

29 G. L. Hobby, K. Meyer, and E. Chaffee, "Observations on the Mechanism of Action of Penicillin," *Proc. Soc. Exp. Biol. & Med.* (1942): 281–85.

30 G. L. Hobby and M. H. Dawson, "Effect of Rate of Growth of Bacteria on Action of Penicillin," *Proc. Soc. Exp. Biol. & Med.* 56 (1944): 181–84.

31 A. Fleming, "The Staphylococci," in *A System of Bacteriology in Relation to Medicine*, Medical Research Council, vol. 2 (London: His Majesty's Station-ery Office, 1929), ch. 1. The techniques used by Fleming in his study of the morphological variants of staphylococci were those in common use in such studies during the 1920s and 1930s. Those of us interested in "bacterial variation" or "bacterial morphology" frequently—indeed almost routinely—set plates aside, at room temperature, after incubation at 37°C, in order to determine if morphological changes would occur. It was not customary or considered important to record the precise temperature or atmospheric conditions in the room. Bacteriology at the time was not a precise science; it was in its infancy. Although the story of penicillin is only a small part of this book by Macfarlane, it is an interesting account. It raises some questions,

however. Macfarlane has relied heavily on the writings of Ronald Hare (see *n*32) who has attempted to re-enact the discovery of penicillin by analyzing and reproducing the conditions that might have existed in the laboratory at St. Mary's Hospital in 1928. That penicillin exerts its growth-inhibitory effects only against multiplying cells was established in my own laboratory in 1941 (*Proc. Soc. Exp. Biol. & Med.* 50 (1942): 281–85) and was subsequently confirmed by many. But growth inhibition and lysis are not the same. The effects of penicillin on growth inhibition and on lysis and death of bacteria have been studied in depth by Alexander Tomasz and by others in recent years, and it seems clear that the mechanism(s) by which penicillin exerts its effect(s) are complex (see *n*34–36). One wonders if Fleming would have reacted to signs of bacteriostatic action as he did to the signs of bacteriolysis had the mold produced only a growth-inhibitory effect.

32 R. Hare, "The Scientific Activities of Alexander Fleming, other than the Discovery of Penicillin," *Medical History* 27 (1983): 347–72.

33 G. Macfarlane, *Alexander Fleming—The Man and the Myth* (Cambridge, Mass.: Harvard University Press, 1984).

34 A. Tomasz, "From Penicillin-Binding Proteins to the Lysis and Death of Bacteria: A 1979 Review," *Revs. of Infect. Diseases* 1, no. 3 (May–June 1979): 434–67.

35 A. Tomasz, "The Mechanism of the Irreversible Antimicrobial Effects of Penicillins: How the Beta-Lactam Antibiotics Kill and Lyse Bacteria," *Ann. Rev. Microbiol.* 33 (1979): 113–37.

36 A. Tomasz, "On the Mechanism of the Irreversible Antimicrobial Effects of Beta-Lactams,' *Phil. Trans. R. Soc. Lond.* B 289 (1980): 303–08.

37 E. B. Chain, "The Early Years of Penicillin Discovery," *TIPS—The Inaugural Issue* 1 (1979): 6–10 (published by Elsevier/North Holland Biomedical Press).

Chapter 2

1 A. Fleming, "On a Remarkable Bacteriolytic Element Found in Tissues and Secretions," *Proc. R. Soc.*, ser. B, 93 (1922): 306–17.

2 A. Fleming, "A Bacteriolytic Ferment Found Normally in Tissues and Secretions," *Lancet* 1 (1929): 217–20.

3 A. Fleming, "Lysozyme (President's Address)," *Proc. Roy. S. Med.* 26 (1932): 71–84.

4 A. Fleming and V. D. Allison, "Further Observations on a Bacteriolytic Element Found in Tissues and Secretions," *Proc. R. Soc.*, ser. B, 94 (1922): 142.

5 A. Fleming and V. D. Allison, "Observations on a Bacteriolytic Substance ('Lysozyme') Found in Tissues and Secretions," *Brit. J. Exp. Pathol.* 3, no. 5 (1922): 252–60.

6 A. Fleming and V. D. Allison, "On the Antibacterial Power of Egg-white," *Lancet* 1 (1924): 1303–07.

7 A. Fleming and V. D. Allison, "On the Specificity of the Protein of Human Tears," *Brit. J. Exp. Pathol.* 6 (1925): 87–90.

8 A. Fleming and V. D. Allison, "On the Effect of Variations of the Salt Content of Blood on its Bactericidal Power *In Vitro* and *In Vivo*," *Brit. J. Exp. Pathol.* 7 (1926): 274–81.

9 A. Fleming and V. D. Allison, "On the Development of Strains of Bacteria Resistant to Lysozyme Action and the Relation of Lysozyme Action to Intracellular Digestion," *Brit. J. Exp. Pathol.* 8 (1927): 214–18.

10 N. E. Goldsworthy and H. W. Florey, "Some Properties of Mucus, with Special Reference to Its Antibacterial Functions," *Brit. J. Exp. Pathol.* 9 (1930): 192–208.

11 H. W. Florey, "The Relative Amounts of Lysozyme Present in the Tissues of Some Mammals," *Brit. J. Exp. Pathol.* 9 (1930): 251–61.

12 Ernst Chain, "The Early Years of the Penicillin Discovery," *TIPS—The Inaugural Issue* 1 (1979): 6–10 (published by Elsevier/North Holland Biomedical Press).

13 E. B. Chain, "A Short History of the Penicillin Discovery from Fleming's Early Observations in 1929 to the Present Time," in *The History of Antibiotics—A Symposium* (Madison, Wis.: American Institute of Pharmacy, 1980).

14 K. Meyer, R. Thompson, J. W. Palmer, and D. Khorazo, "The Purification and Properties of Lysozyme," *J. Biol. Chem.* 113 (1936): 303–09.

15 K. Meyer, R. Thompson, J. W. Palmer, and D. Khorazo, "On the Mechanism of Lysozyme Action," *J. Biol. Chem.* 113 (1936): 479–86.

16 K. Meyer and J. W. Palmer, "The Enzyme-Hydrolyzed Hyaluronic Acid (a Mucopolysaccharide) Present in Vitreous Humour and Synovial Fluid," *J. Biol. Chem.* 114 (1936): 689–703.

17 E. A. H. Roberts, "The Preparation and Properties of Purified Egg-white Lysozyme," *Quart. J. Exp. Physiol.* 27 (1937): 89–98.

18 E. P. Abraham, "Some Properties of Egg-white Lysozyme," *Biochem. J.* 33 (1939): 622–30.

19 L. A. Epstein and E. B. Chain, "Some Observations on the Preparation and Properties of the Substrate of Lysozyme," *Brit. J. Exp. Pathol.* 21 (1940): 339–55.

20 E. B. Chain and H. W. Florey, "Penicillin," *Endeavour.* 3, no. 9 (January 1944).

21 M. R. K. Salton, "Properties of Lysozyme and Its Action on Microorganisms," *Bact. Rev.* 21 (1957): 82–99.

22 K. Meyer and M. M. Rapport, "The Mucopolysaccharides of the Ground Substance of Connective Tissue," *Science* 113, no. 2943 (1951): 596–99.

23 Giuliano Quintarelli, ed., *The Chemical Physiology of Mucopolysaccharides* (Boston: Little Brown & Co., 1968).

24 D. Kaplan, "Presentation of Academy Medal to Karl Meyer, M.D. (Stated Meeting of the New York Academy of Medicine, April 9, 1981)," *Bull. New York Acad. Med.* 57, no. 8 (October 1981): 667–73.

25 E. P. Abraham, "Howard Walter Florey—Baron Florey of Adelaide and Marston, 1898–1968," *Biographical Memoirs of Fellows of the Royal Society,*

vol. 17, November 1971.

26 H. W. Florey, E. Chain, N. G. Heatley, M. A. Jennings, A. G. Saunders,
    E. P. Abraham, and M. E. Florey, *Antibiotics: A Survey of Penicillin, Strep-
    tomycin, and Other Antimicrobial Substances from Fungi, Actinomycetes,
    Bacteria and Plants* (London: Oxford University Press, 1949), pp. 636–37.

27 H. W. Florey, Archives, Royal Society (London), 98 HF 36.17.1. Application
    to the Rockefeller Foundation for Research Grant, dated 11/20/39.

28 R. Dubos, "Studies on a Bactericidal Agent Extracted from a Soil Bacillus.
    I. Preparation of the Agent. Its In Vitro Activity," *J. Exp. Med.* 70 (1939):
    1–10.

29 R. Dubos, "Studies on a Bactericidal Agent Extracted from a Soil Bacillus.
    II. Protective Effect of the Bactericidal Agent against Experimental Pneu-
    mococcal Infections in Mice," *J. Exp. Med.* 70 (1939): 11–17.

30 F. Duran-Reynals, "The Effect of Extracts of Certain Organs from Normal
    and Immunized Animals on the Infecting Power of Vaccine Virus," *J. Exp.
    Med.* 50 (1929): 327–40.

31 D. C. Hoffman and F. Duran-Reynals, "The Influence of Testicle Extract on
    the Intradermal Spread of Injected Fluids and Particles," *J. Exp. Med.* 53
    (1931): 387–98.

32 F. Duran-Reynals, "Studies on a Certain Spreading Factor Existing in
    Bacteria and Its Significance for Bacterial Invasiveness," *J. Exp. Med.* 58
    (1933): 161–82.

33 K. Meyer and J. W. Palmer, "On Glycoproteins. II. The Polysaccharides of
    Vitreous Humour and Umbilical Cord," *J. Biol. Chem.* 114 (1936): 689–703.

34 F. E. Kendall, M. Heidelberger, and M. H. Dawson, "A Serologically
    Inactive Polysaccharide Elaborated by Mucoid Strains of Group A Strepto-
    cocci," *J. Biol. Chem.* 118 (1937): 61–69.

35 K. Meyer, G. L. Hobby, E. Chaffee, and M. H. Dawson, "The Hydrolysis of
    Hyaluronic Acid by Bacterial Enzymes," *J. Exp. Med.* 71, no. 2 (1940):
    137–46.

36 K. Meyer, G. L. Hobby, E. Chaffee, and M. H. Dawson, "The Relationship
    between 'Spreading Factor' and Hyaluronidase," *Proc. Soc. Exp. Biol. &
    Med.* 44 (1940): 294–96.

37 G. L. Hobby, M. H. Dawson, K. Meyer, and E. Chaffee, "The Relationship
    between Spreading Factor and Hyaluronidase," *J. Exp. Med.* 73, no. 1 (1941):
    109–23.

38 K. Meyer, E. Chaffee, G. L. Hobby, and M. H. Dawson, "Hyaluronidases of
    Bacterial and Animal Origin," *J. Exp. Med.* 73, no. 3 (1941): 309–26.

39 T. D. Day, "The Permeability of Interstitial Connective Tissue and the Nature
    of the Interfibrillary Substance," *J. Physiol.* 117 (1952): 1–8.

40 T. D. Day, "The Nature of the Spreading Effects Occasioned by Hyaluroni-
    dase in Subcutaneous Connection Tissue," *Brit. J. Exp. Pathol.* 38 (1957):
    326–28.

41 E. Chain and E. S. Duthie, "Identity of Hyaluronidase and Spreading
    Factor," *Brit. J. Exp. Pathol.* 21 (1940): 324–38.

42 John Toland, *Adolf Hitler*, 2 vols. (Garden City, N.Y.: Doubleday & Co.,

1976).

43 G. Macfarlane, *Howard Florey—The Making of a Great Scientist* (Oxford: Oxford University Press, 1979), p. 299.

44 H. W. Florey and E. Mellanby, Correspondence, June–September 1939, Medical Research Council Archives, London, file No. 1752, vol. 1.

Chapter 3

1 G. B. Shaw, *The Doctor's Dilemma* (New York: Dodd Mead and Co., 1906, reprinted 1941).

2 L. Colebrook, "Almroth Edward Wright, 1861–1947," *Obituary Notices of Fellows of the Royal Society* 6 (November 1948): 296–314.

3 L. Colebrook, *Almroth Wright—Provocative Doctor and Thinker* (London: William Heinemann—Medical Books, Ltd., 1954).

4 P. H. Hiss, Jr. and H. Zinsser, *A Text-Book of Bacteriology*, 3d ed. (New York and London: D. Appleton and Co., 1916).

5 R. Hare, *The Birth of Penicillin* (Oxford: George Allen and Unwin, 1970).

6 F. Griffith, "The Influence of Immune Serum on the Biological Properties of Pneumococci," in *Reports on Public Health and Medical Subjects. No. 18. Bacteriological Studies* (London: H.M.S.O., 1923), pp. 1–13.

7 F. Griffith, "The Significance of Pneumococcal Types," *J. Hyg.* 27 (1928): 113–59. Neufeld and Händel in 1909 first demonstrated the existence of anti-genically different types of pneumococci. Somewhat later, Dochez, Avery, and their associates studied a large number of strains of pneumococci, mostly from patients with lobar pneumonia, and noted that there were three well-differentiated types (I, II, and III). It is now known that more than eighty serologically different types exist. The smooth (S) to rough (R) variation in pneumococci, studied by Griffith, is associated with the loss of the organism's characteristic capsule and with it the polysaccharide antigen that confers type-specificity. Griffith's conversion of a smooth strain of pneumococci belonging to one specific antigenic type, through the corresponding rough variant to a smooth strain of another antigenic type was the first instance of a transfor-mation of one normal bacterial cell into another under experimental con-ditions. These observations have been abundantly confirmed, and their broad significance is now well recognized. (See F. Neufeld and F. Händel, *Arb. Reighsgesundh Amt.* 34 [1909]: 293; A. R. Dochez et al., *J. Exper. Med.* 21 [1915], 114; ibid. 61 [1913], 727; ibid. 30 [1919], 179; R. Austrian and C. M. MacLeod, *J. Exper. Med.* 89 [1949], 451; R. M. Bracco, M. R. Knauss, A. S. Roe, and C. M. MacLeod, *J. Exper. Med.* 106 [1957]: 247.)

8 O. T. Avery, C. MacLeod, and M. McCarty, "Studies on the Chemical Nature of the Substance Inducing Transformation of Pneumococcal Types," *J. Exp. Med.* 79 (1944): 137–58.

9 M. McCarty and O. Avery, "Studies on the Chemical Nature of the Substance Inducing Transformation of Pneumococcal Types. II. Effect of Deoxyribonuclease on the Biological Activity of the Transforming Sub-stance," *J. Exp. Med.* 83 (1946): 89–96.

10 M. McCarty and O. Avery, "Studies on the Chemical Nature of the Substance Inducing Transformation of Pneumococcal Types. III. An Improved Method for the Isolation of the Transforming Substance and Its Application to Pneumococcus Types II, III, and IV," *J. Exp. Med.* 83 (1946): 97–104.

11 K. Baerthlein, "Über bacterielle variebilitat, insebesondere sogenannte bakterienmutationen," *Cent. f. Bakt.* I.0. 81 (1918): 369.

12 J. A. Arkwright, "Variation in Bacteria in Relation to Agglutination Both by Salts and by Specific Serum," *J. Pathol. Bacteriol.* 24 (1921): 36–.

13 M. H. Dawson, "The Transformation of Pneumococcal Types I and III," *J. Exp. Med.* 51 (1930): 99–122, 123–47.

14 M. H. Dawson and R. Sia, "The In Vitro Transformation of Pneumococcal Types I and II," *J. Exp. Med.* 54 (1931): 681–99, 701–10.

15 R. J. Dubos, *The Professor, The Institute, and DNA* (New York: Rockefeller University Press, 1976).

16 F. Griffith, "The Serological Classification of Streptococcus Pyogenes," *J. Hygiene, Cambridge* 34 (1934): 542–84.

17 R. C. Lancefield, "The Antigenic Complex of Streptococcus Hemolyticus. I. Demonstration of a Type-Specific Substance in Extracts of Streptococcus Hemolyticus," *J. Exp. Med.* 47 (1928): 91–103.

18 R. C. Lancefield, "The Antigenic Complex of Streptococcus Hemolyticus. II. Chemical and Immunological Properties of the Protein Fractions," *J. Exp. Med.* 47 (1928): 469–80.

19 R. C. Lancefield, "The Antigenic Complex of Streptococcus Hemolyticus. III. Chemical and Immunological Properties of the Species-Specific Substance," *J. Exp. Med.* 47 (1928): 481–91.

20 R. C. Lancefield, "The Antigenic Complex of Streptococcus Hemolyticus. IV. Anaphylaxis with Two Non-Type-Specific Fractions," *J. Exp. Med.* 47 (1928): 843–55.

21 R. C. Lancefield, "The Antigenic Complex of Streptococcus Hemolyticus. V. Anaphylaxis with the Type-Specific Substance," *J. Exp. Med.* 47 (1928): 857–75.

22 R. C. Lancefield, "A Serological Differentiation of Human and Other Groups of Hemolytic Streptococci," *J. Exp. Med.* 57 (1933): 571–95.

23 R. C. Lancefield, "Specific Relationship of Cell Composition to Biological Activity of Hemolytic Streptococci," *Harvey Lectures*, Series B 36 (1940–41): 25.

24 Sir Graham S. Wilson and Sir A. Ashley Miles, eds., *Topley and Wilson's Principles of Bacteriology and Immunology*. 5th ed., 2 vols. (Baltimore: Williams and Wilkins Co., 1964), ch. 23, p. 704.

25 P. Guttmann and P. Ehrlich, "Über die Wirkung des Methylenblau bei Malaria," *Berl. Klin. Wochr.* (1981). Translated by the editors of *The Collected Papers of Paul Ehrlich*, ed. F. Himmelweit with the assistance of M. Marquardt, under the editorial direction of Sir Henry Dale, vol. 3 (London and New York: Pergamon Press, 1960), pp. 1–8, 9–20.

26 P. Ehrlich, *The Collected Papers of Paul Ehrlich*, ed. F. Himmelweit with the assistance of M. Marquardt, under the editorial direction of Sir Henry Dale,

vol. 3, pp. 1–8, 53–63, 282–309.

27 H. W. Thomas and A. Breinl, *Report on Trypanosomes, Trypanosomiasis, and Sleeping Sickness* (London: Cleveland Press, 1905).

28 P. Ehrlich, *Die Behandlung der Syphilis mit dem Ehrlichschen Präparat 606*, Verh 82, Vers. Ges dtsch. Naturf. Ärzt, 1911, in *The Collected Papers of Paul Ehrlich*, pp. 240–46.

29 P. Ehrlich and A. Berthheim, "Über das salzsäure 3.3'-diamino-4.4' dioxylarsenobenzol und seine nächsten Verwandten," *Ber. dtsch. chem. Ges.* 45 (1912): 756; and *The Collected Papers of Paul Ehrlich*, pp. 402–11.

30 P. Ehrlich and S. Hata, *The Experimental Chemotherapy of Spirilloses* (London: Ribman, 1911).

31 P. Ehrlich, "On Partial Functions of the Cell," trans. by the Editors from Les Prix Nobel, Stockholm, 1909, in *The Collected Papers of Paul Ehrlich*, pp. 183–94.

32 P. Ehrlich, "Chemotherapy." Reprinted with amendments by the ed. from Proc. 17th Int. Congr. Med. (1913) 1914, in *The Collected Papers of Paul Ehrlich*, pp. 505–18.

33 G. Domagk, "Ein Beitrag zur Chemotherapie der bakteriellen Infektionen," *Deutsche. med. Wchnschr.* 61 (1935): 250.

34 E. H. Northey, *The Sulfonamides and Allied Compounds* (New York: Reinhold Publishing Corp., 1948), pp. v–vii.

35 P. Gelmo, "Über Sulfamide der p-Amidobenzolsulfonsäure," *J. Prakt. Chem.* 77 (1908): 369.

36 H. Hörlein, O. Dressel, and H. Kothe, *Deutsches Reichpatent* (1909–10): 226, 230, 235, 239, 240, 594, 775.

37 P. Eisenberg, "Über die Wirkung von Farbstoffen auf Bakterien," *Zentralbl f. Bakt. Abt.* 71 (1913): 420.

38 M. Heidelberger and W. A. Jacobs, "Azo Dyes Derived from Hydrocupreine and Hydrocupreidine," *J. Am. Chem. Soc.* 41 (1919): 2131.

39 P. H. Long and E. A. Bliss, *The Clinical and Experimental Use of Sulfanilamide, Sulfapyridine, and Allied Compounds* (New York: MacMillan Co., 1939), pp. 1–13.

40 H. Hörlein, O. Dressel, and H. Kothe, *Deutsches Reichpatent* (1909–10): 226, 230, 235, 239, 240, 594, 775.

41 W. Schulemann, F. Schönhöffer, and A. Wengler, "Synthese des plasmochin," *Klin. Wschr.* 11 (1932): 381.

42 H. Mauss and F. Mietzsch, "Atebrin, ein neues Heilmittel gegen Malaria," *Klin. Wschr.* 12 (1933): 1276.

43 J. Morgenroth and R. Levy, "Chemotherapie der Pneumokokkeninfektionen," *Berl. Klin. Wschr.* 48 (1911): 1979.

44 L. Colebrook, "Gerhard Domagk 1865–1964," *Biographical Memoirs of Fellows of the Royal Society* 10 (1964): 39–50.

45 Ph. Klee and H. Römer, "Prontosil bei streptokokkenertrankungen," *Deutsche. med. Wchnschr.* 61 (1935): 253.

46 T. H. Schreus, "Chemotherapiedes erysipels und anderer infektionen mit prontosil," *Deutsche. Med. Wchnschr.* 61 (1935): 255.

47 E. Anselm, "Unsere erfahrungen mit prontosil bei puerperalfieber," *Deutsche. Med. Wchnschr.* 61 (1935): 264.

48 Dr. Foerster, "Case Report on Streptozon," *Zentralbl. f. Haut u. Geschlechtskr.* 45 (1933): 549.

49 C. Levaditi and A. Vaisman, "Chimotherapie. Action curative et preventive du chlorhydrate de 4'-sulfamido-2,4-diamino azobenzene dans l'infection streptococcique experimentale," *Compt. rend. Acad. d. Sc.* 200 (1935): 1694–95.

50 F. Nitti and D. Bovet, *Abstr. Compt. rend. Soc. de Biol.* 119 (1935): 1277.

51 J. Tréfouël, Mme. J. Tréfouël, F. Nitti, and D. Bovet, "Activité du p-aminophénylsulfamide sur les infections streptococciques expérimentales de la souris et du lapin," *Compt. rend. Soc. de Biol.* 120 (1936): 756.

52 L. Colebrook and Maeve Kenny, "Treatment of Human Puerperal Infections, and of Experimental Infections in Mice, with Prontosil," *Lancet* 1 (June 6, 1936): 1279–86.

53 G. A. H. Buttle, W. H. Gray, and D. Stephenson, "Protection of Mice against Streptococcal and Other Infections by Para-Aminobenzene Sulfonamide and Related Substances," *Lancet* 1 (1936): 1286.

54 A. T. Fuller, "Is Para-Aminobenzenesulfonamide the Active Agent in Prontosil Therapy?" *Lancet* 1 (1935): 194.

55 L. Colebrook and A. W. Purdie, "Treatment of 106 Cases of Puerperal Fever by Sulfanilamide," *Lancet* 2 (November 27, 1939): 1237–42.

56 L. G. Matthews, *History of Pharmacy in Britain* (Edinburgh and London: E. & S. Livingstone, 1962), pp. 228, 231, 343–44, 348.

57 L. H. Sophian, D. L. Piper, and G. H. Schneller, *The Sulfapyrimidines* (New York: by the Press of A. Colish, 1952). Sulfanilamide consists of a benzene ring attached at one end to a sulfonamide group ($-SO_2NH_2$) and at the other end, in the para position, to an amino ($-NH_2$) group. If the amino group is replaced by another group or moved to the meta or ortho position on the benzene ring, there is a decrease in activity. Thus the development of active derivatives of sulfanilamide centered on synthesizing compounds in which the substituents were attached to the sulfonamide group.

58 L. E. H. Whitby, "Chemotherapy of Pneumococcal and Other Infections with 2-(Para-Aminobenzenesulfonamido) Pyridine," *Lancet* 1 (1938): 1210. The pyridine analog of sulfapyridine and sulfathiazole was synthesized in 1940 (R. O. Roblin, Jr., J. H. Williams, P. S. Winnek, and J. P. English, "Chemotherapy: II. Some Sulfanilamide Heterocytes," *J. Amer. Chem. Soc.* 62 [1940]: 2002) and was designated sulfadiazine. Less toxic than the earlier compounds, it was highly effective against a wide range of bacterial infections (M. Finland, E. Strauss, and O. L. Peterson, "Sulfadiazine: Therapeutic and Toxic Effects on Four Hundred and Forty-six Patients," *J. Amer. Med. Ass.* 116 [June 14, 1941]: 2641–47). Now, more than forty years later, sulfadiazine remains a widely used and effective chemotherapeutic drug (R. G. Petersdorf, "Landmark Perspective on Sulfadiazine," *J. Amer. Med. Assoc.* 251, no. 11 [March 16, 1984]: 1475–76).

59 R. Emmerich and O. Low, "Bakteriolytische Enzyme als Ursache der

erworbenen Immunität und die Heilung von Infektionskrankheiten durch dieselben," *Z. Hyg. InfektKr.* 31 (1899): 1.

60 E. Bliss and P. H. Long, "Activation of 'Prontosil Solution' in Vitro by Reduction with Cystine Hydrochloride," *Bull. Johns Hopkins Hosp.* 60 (February 1937): 149–53.

61 P. H. Long and E. A. Bliss, "Para-Amino-Benzene Sulfonamide and Its Derivatives," *J. Amer. Med. Assoc.* 108 (1937): 32.

62 P. H. Long and E. A. Bliss, "The Use of Para Amino Benzene Sulfonamide (Sulfanilamide) or Its Derivatives in the Treatment of Infections Due to Beta Hemolytic Streptococci, Pneumococci, and Meningococci," *South. Med. J.* 30 (1937): 479.

63 E. K. Marshall, Jr., K. Emerson, Jr., and W. C. Cutting, "Para-Aminobenzene Sulfonamide," *J. Amer. Med. Assoc.* 108 (1937): 953.

64 E. K. Marshall, Jr., K. Emerson, Jr., and W. C. Cutting, "The Distribution of Sulfanilamide in the Organisms," *J. Pharmacol. and Exper. Therap.* 61 (1936): 196.

65 E. K. Marshall, Jr., K. Emerson, Jr., and W. C. Cutting, "The Toxicity of Sulfanilamide," *J. Amer. Med. Assoc.* 110 (1938): 252–57.

66 D. D. Woods, "The Relation of Para-Aminobenzoic Acid to the Mechanism of Action of Sulfanilamide," *Brit. J. Exp. Pathol.* 21 (April 1940): 74.

67 P. Fildes, "A Rational Approach to Research in Chemotherapy," *Lancet* 1 (May 25, 1940): 955.

68 J. Landy, N. W. Larkum, E. J. Oswald, and F. Streighthoff, "Increased Synthesis of Para-Amino-Benzoic Acid Associated with the Development of Sulfonamide Resistance in *Staphylococcus aureus*," *Science* 97 (1943): 265.

69 M. Welsch, *Phénomènes d'antibiose chez les Actinomycetes*, Université de Liège: Thèse D'Agrégation de l'Enseignement Supérieur (Imprimerie J. Duculot, Gembloux, 1947).

70 S. A. Waksman, *Microbial Antagonisms and Antibiotic Substances* (New York: Commonwealth Fund, 1945).

71 R. J. Dubos, "Studies on a Bactericidal Agent Extracted from a Soil Bacillus. I. Preparation of the Agent. Its Activity *In Vitro*. II. Protective Effect of the Bactericidal Agent Against Experimental Pneumococcus Infections in Mice," *J. Exp. Med.* 70 (1939): 1–10, 11–17.

72 R. J. Dubos and R. D. Hotchkiss, "The Production of Bactericidal Substances by Aerobic Sporulating Bacilli," *J. Exp. Med.* 73 (1941): 629–40.

73 A. Fleming, "A Comparison of the Activities of Antiseptics on Bacteria and on Leucocytes," *Proc. Roy. Soc. of London*, ser. B, 96 (1924): 171–80.

74 A. Fleming, "Antiseptics. The Lister Memorial Lecture," *Chemistry and Industry* 1, no. 3 (January 20, 1945): 18–23.

75 L. Colebrook, "Alexander Fleming 1881–1955," *Biographical Memoirs of Fellows of the Royal Society* 2 (1956): 117–27.

76 A. Fleming and I. H. Maclean, "On the Occurrence of Influenzae Bacilli in the Mouths of Normal People," *Brit. J. Exp. Pathol.* 11 (1930): 127–34.

77 A. Fleming, "On the Specific Antibacterial Properties of Penicillin and Potassium Tellurite—Incorporating a Method of Demonstrating Some

Bacterial Antagonisms," *J. Pathol. Bacteriol.* 35 (1932): 831–42.

78 E. B. Chain, "The Early Years of the Penicillin Discovery," *TIPS—The Inaugural Issue* (1979): 6–10; published by Elsevier/North Holland Biomedical Press.

79 P. W. Clutterbuck, R. Lovell, and H. Raistrick, "Studies in the Biochemistry of Microorganisms. No. 26. The Formation from Glucose by Members of the *Penicillium chrysogenum* Series of a Pigment, an Alkali-Soluble Protein, and Penicillin—The Antibacterial Substance of Fleming," *Biochem. J.* 26 (1932): 1907–18.

80 A. Fleming, "On the Antibacterial Action of Cultures of Penicillium, with Special Reference to Their Use in the Isolation of *H. influenzae*," *Brit. J. Exp. Pathol.* 10, no. 3 (1929): 226–36.

81 Stuart Craddock, Laboratory Notebook of S. Craddock (St. Mary's Hospital, Paddington) for September 5, 1928, to March 1930. In: The Fleming Papers, The British Library, London 56224 (Vol. CXIX).

82 Peter Regna, 1980, personal communication.

83 A. J. P. Martin and R. L. M. Synge, "A New Form of Chromatogram Employing Two Liquid Phases. (1) A Theory of Chromatography. (2) Application to the Micro-Determination of the High Monoamino Acids in Proteins," *Biochem. J.* 35 (1941): 1358–68.

84 R. Consden, A. H. Gordon, and A. J. P. Martin, "Qualitative Analysis of Proteins: A Partition Chromatographic Method Using Paper," *Biochem. J.* 38 (1944): 224–32.

85 R. R. Goodall and A. A. Levi, "A Micro-Chromatographic Method for the Detection and Approximate Determination of the Different Penicillins in a Mixture," *Analyst.* 72 (1947): 277–88.

86 K. B. Raper, "The Penicillin Saga Remembered," *ASM News* 44, no. 12 (Washington, D.C.: American Society for Microbiology, 1944): 645–54.

87 Charles Thom, "Mycology Presents Penicillin," *Mycologia* 37, no. 4 (1945): 460–75.

88 Charles Thom, "Molds, Mutants, and Monographers," *Mycologia* 44, no. 1 (1952): 61–85.

89 R. D. Reid, "Some Observations on the Ability of a Mold, or Its Metabolic Products, to Inhibit Bacterial Growth," *J. Bacteriol.* (Proc.) 27 (1934): 28.

90 R. D. Reid, "Some Properties of a Bacterial-Inhibitory Substance Produced by a Mold," *J. Bacteriol.* 29 (1935): 215–21.

91 R. D. Reid, personal communication, 1950, to Dr. Gladys L. Hobby.

92 R. Hare, *The Birth of Penicillin* (Oxford: George Allen and Unwin, 1970), pp. 103–04, 169.

93 A. Fleming, Archives, British Library (London), 789E. Collections 56156 through 56166, for the years 1924 through 1932.

94 S. Bornstein, "Action of Penicillin on Enterococci and Other Streptococci," *J. Bacteriol.* 39 (April 1940): 383–87.

95 C. G. Paine, Unpublished personal communication to Professor H. W. Florey. See H. W. Florey, E. Chain, N. G. Heatley, M. A. Jennings, A. G. Sanders, E. P. Abraham, and M. E. Florey, *Antibiotics—A Survey of*

Penicillin, Streptomycin and Other Antimicrobial Substances from Fungi, Actinomycetes, Bacteria and Plants (London, New York, and Toronto: Oxford University Press, 1949), p. 634.

96 R. Hare, *The Birth of Penicillin and the Disarming of Microbes* (London: George Allen and Unwin, 1970), pp. 105–06.

97 L. J. Ludovici, *Fleming—Discover of Penicillin* (London: Andrew Dakers, 1952), pp. 147–48.

98 H. W. Florey, Personal communication to Dr. Walsh McDermott.

99 A. Fleming, "Streptococcal Meningitis Treated with Penicillin—Measurement of Bactericidal Power of Blood and Cerebrospinal Fluid," *Lancet* 2 (October 9, 1943): 434–45.

100 A. Fleming, "The Staphylococci," in *A System of Bacteriology in Relation to Medicine* 2, The Medical Research Council (London: His Majesty's Stationery Office, 1929).

101 *Reported Incidence of Selected Notifiable Diseases: United States, Each Division and State, 1920–1950. Vital Statistics—Special Reports/National Summaries* 37, no. 9 (U.S. Department of Health, Education, and Welfare, Public Health Service, National Office of Vital Statistics, June 15, 1953).

102 R. L. Cecil, ed., *A Textbook of Medicine by American Authors* (Philadelphia and London: W. B. Saunders, 1943).

103 H. Dowling, *Fighting Infection: Conquests of the Twentieth Century*, A Commonwealth Fund Book (Cambridge, Mass., and London: Harvard University Press, 1977).

104 *Annual Summary, 1979. Reported Morbidity and Mortality in the United States.* MMWR Weekly Report, vol. 28, no. 54 (U.S. Department of Health and Human Services, Public Health Service, September 1980).

105 R. H. Shryock, *American Medical Research, Past and Present* (New York: Commonwealth Fund, 1947).

106 Sir Graham S. Wilson and Sir A. Ashley Miles, eds., *Topley and Wilson's Principles of Bacteriology and Immunology*, 5th ed., vol. 1 (Baltimore, Md.: Williams & Wilkins Co., 1964).

107 Fleming published at least five reports in 1938–40 on the action of sulfanilamide and sulfapyridine (*Lancet* 2 [1938]: 74; *Lancet* 2 [1938]: 564; *Lancet* 1 [1939]: 562 [with I. H. Maclean and K. B. Rogers]; Tr. Roy. Soc. Med. 32 [1939]: 911; Tr. Med. Soc. Lond. 62 [1939]: 19–43; J. Path. Bact. 50, no. 1 [1940]: 69). Once his attention was directed to the chemotherapeutic potential of penicillin by the work of Florey and his team at Oxford, Fleming seemed to lose interest in pursuing study of the sulfonamides. In a little-known book on Fleming and the discovery of penicillin published in 1952, L. J. Ludovici—who obviously knew Fleming and his close associates well—implies that he had talked with Fleming extensively on this subject and indicates that Fleming was greatly impressed (1) by the immediacy of the increased antistreptococcal action of serum after intravenous injection of animals with sulfanilamide; (2) by the significance of the number of infecting organisms (or size of inoculum) on the degree of activity demonstrable with sulfanilamide; and (3) by the fact that in vivo as in vitro, these substances

were bacteriostatic but not bactericidal. To him this implied need for immune factors within the host to remove the residual suppressed but still viable infecting organisms. For a time at least, Fleming advocated the use of vaccines in combination with the sulfonamides.

Chapter 4

1 D. Wilson, *In Search of Penicillin* (New York: Alfred A. Knopf, 1976), p. 3.

2 E. B. Chain, "The Early Years of the Penicillin Discovery," *TIPS—The Inaugural Issue* (Elsevier/North Holland Biomedical Press). Anaphylactic reactions in guinea pigs, or what was assumed to be evidence of anaphylaxis (or hypersensitivity), were observed early—in my own laboratories and by Dr. Geoffrey Rake soon after. They were atypical reactions, they did not occur routinely, and they did not prevent our demonstrating the chemotherapeutic effects of penicillin against *Clostridial* infections in these animals. More important though, there were no deaths due to administration of the drug to more than ten thousand patients during the course of Dr. Chester Keefer's (O.S.R.D.) study to evaluate the therapeutic efficacy of penicillin (see chapter 8). Although today hypersensitivity to penicillin is manifest in 3 to 5 percent of those treated with the drug, and an occasional one of these may die of anaphylaxis, severe hypersensitivity reactions in no way interfered with the early development of penicillin.

3 E. P. Abraham, "Ernst Boris Chain, 1906–1979," *Biographical Memoirs of Fellows of the Royal Society*, vol. 29, November 1983.

4 E. B. Chain, "Thirty Years of Penicillin Therapy," *Proc. R. Soc. Lond. B.* 179 (1971): 293–319.

5 E. Chain, H. W. Florey, A. D. Gardner, N. G. Heatley, M. A. Jennings, J. Orr-Ewing, and A. G. Sanders, "Penicillin as a Chemotherapeutic Agent," *Lancet* 2 (August 24, 1940): 226–28.

6 H. W. Florey, E. Chain, N. G. Heatley, M. A. Jennings, A. G. Sanders, E. P. Abraham, and M. E. Florey, *Antibiotics—A Survey of Penicillin, Streptomycin, and Other Anti-Microbial Substances from Fungi, Actinomycetes, Bacteria, and Plants*, 2 vols. (London: Oxford University Press, 1949).

7 Florey Archives, Library of the Royal Society, London, 98 HF 34, Experiment I on "Curative Effect of Penicillin, *In Vivo*," Streptococci, dated May 25, 1940; later experiment dated July 11, 1940.

8 E. P. Abraham, E. Chain, C. M. Fletcher, H. W. Florey, A. D. Gardner, N. G. Heatley, and M. A. Jennings, "Further Observations on Penicillin," *Lancet* 2 (August 16, 1941): 177–88.

9 R. B. Fosdick, *The Story of the Rockefeller Foundation* (New York: Harper and Brothers, 1953), pp. 165–66 and p. 274.

10 Gladys L. Hobby, Unpublished materials, personal laboratory records.

11 M. H. Dawson and G. L. Hobby, "The Clinical Use of Penicillin—Observations in One Hundred Cases," *J. Amer. Med. Assoc.* 124 (March 4, 1944): 611–22.

12 M. H. Dawson and T. H. Hunter, "The Treatment of Subacute Bacterial Endocarditis with Penicillin—Results in Twenty Cases," *J. Amer. Med. Assoc.* 127 (January 20, 1945): 129–37.

13 B. Sokoloff, *The Story of Penicillin* (Chicago and New York: Ziff-Davis Publishing Co., 1945), p. 2.

14 M. H. Dawson, G. L. Hobby, K. Meyer, and E. Chaffee, "Penicillin as a Chemotherapeutic Agent," *J. Clin. Invest.* 20, Abstr. (1941): 434.

15 M. H. Dawson, G. L. Hobby, K. Meyer, and E. Chaffee, "Penicillin as a Chemotherapeutic Agent," presented at the Annual Meeting of the American Society of Clinical Investigation, May 5, 1941.

16 S. Spencer, "Germ Killer Found in Common Mold," *Philadelphia Evening Bulletin*, May 5, 1941.

17 W. L. Lawrence, "Giant Germicide Yielded by Mold," *New York Times*, May 6, 1941, p. 23.

18 W. Herrell and D. Heilman, Personal communication to Dr. Gladys L. Hobby, May 1979.

19 Pfizer Inc., "Historical Records Compiled by George B. Stone, Covering the Period 1920 to 1950," vol. 3, 1941. Unpublished.

20 J. D. Ratcliff, *Yellow Magic—The Story of Penicillin* (New York: Random House, 1945).

21 A. D. Gardner, "Morphological Effects of Penicillin on Bacteria," *Nature* 146 (December 28, 1940): 837–38.

22 Symposium, "Significance of Subminimal Inhibitory Concentrations of Antibiotics," *Review of Infectious Diseases* 1, no. 5 (September–October 1979): 780–879.

23 M. E. Delafield, E. Straker, and W. W. C. Topley, "Antibiotic Snuffs," *Brit. Med. J.* (January 2, 1941): 145–50.

24 E. Chain, "Mode of Action of Chemotherapeutic Agents," *Lancet* 2 (December 20, 1941): 762. (In Soc. Proc.).

25 E. P. Abraham, "Mode of Action of Chemotherapeutic Agents," *Lancet* 2 (December 20, 1941): 762. (In Soc. Proc.).

26 Gladys L. Hobby, K. Meyer, M. H. Dawson, and E. Chaffee, "The Antibacterial Action of Penicillin," *Proc. Soc. Amer. Bact.* (December 1941): 11–12.

27 G. L. Hobby, K. Meyer, and E. Chaffee, "Activity of Penicillin *In Vitro*," *Proc. Soc. Expt. Biol. and Med.* 50 (1942): 277–80.

28 Gladys L. Hobby, K. Meyer, and E. Chaffee, "Observations on the Mechanism of Action of Penicillin," *Proc. Soc. Expt. Biol. and Med.* 50 (1942): 281–85.

29 Gladys L. Hobby, K. Meyer, and E. Chaffee, "Chemotherapeutic Activity of Penicillin," *Proc. Soc. Expt. Biol. and Med.* 50 (1942): 285–88.

30 R. Lopez, C. Ronda-Lain, A. Tapia, S. Waksman, and A. Tomasz, "Suppression of the Lytic and Bactericidal Effects of Cell Wall Inhibitory Antibiotics," *Antimicrobial Agents and Chemotherapy* 10 (1976): 697–706.

31 K. Meyer, E. Chaffee, Gladys L. Hobby, M. H. Dawson, E. Schwenk, and G. Fleischer, "On Penicillin," *Science* 96, no. 2479 (July 3, 1942): 20–21.

274

32   J. F. Fulton, Personal diary, vol. XVII, 1940. Historical Library, Yale University Medical Library, New Haven, Conn. Letter dated April 30, 1947 from Dr. John F. Fulton to Mr. Milton Silverman, science editor, *San Francisco Chronicle*: "The Yale Faculty on 11 June 1940, just after the fall of France, sent a cable to the Oxford faculty saying that they would be glad to look after children of the Oxford faculty members for the duration. Invasion seemed imminent at that time and the Oxford faculty sent a group of 125 children and 23 expectant mothers as a group to New Haven. They arrived in the middle of July and after a week spent at the Children's Center... they were eventually assigned to individual families here in New Haven. We did not know that the Florey children were in the group until they landed, and when we discovered it, we asked the local committee to assign them to us which was done.... The English people at that time were generally alarmed over the course of the war and many felt that their children should, if possible, be spared the hazards of invasion and occupation." (From The Yale University Library, the Fulton Collection.)

33   C. S. Keefer, "Penicillin: A Wartime Achievement," in *Advances in Military Medicine* (Boston: Little, Brown & Co., 1948), chapter 52, pp. 717–22.

34   N. E. Goldsworthy and H. W. Florey, "Some Properties of Mucus with Special Reference to Its Antibacterial Functions," *Brit. J. Exp. Pathol.* 9 (1930): 192–208.

35   N. G. Heatley, "A Method for the Assay of Penicillin," *Biochem. J.* 38, no. 1 (1944): 61–65.

36   L. P. Garrod and N. G. Heatley, "Bacteriological Methods in Connection with Penicillin Treatment," *Brit. J. Surg.* 32, Special Penicillin Issue (July 1944): 117–24.

37   K. Raper, "The Penicillin Saga Remembered," *ASM News* 44, no. 12 (1978): 645–53. The first penicillin standard was prepared in England and had an established potency of 42 Oxford units per milligram. Eventually it was established that the Oxford unit was equal to the biological activity represented by 0.6 microgram of pure sodium penicillin G (benzyl penicillin). An international conference in London in 1944 established this measurement as the international unit. The first sample of the international standard was prepared from a pool of 3-gram samples supplied by each of the major penicillin producers. These were pooled and crystallized as one lot, and its physical characteristics were determined. On this basis, it was established that 1 milligram of crystalline penicillin G contains the equivalent of 1667 units of penicillin activity. The international standard was prepared at the Northern Regional Research Laboratory in Peoria by Dr. Frank Stodola.

38   N. G. Heatley, "In Memoriam, H. W. Florey: An Episode," *J. Gen. Microbiol.* 61 (1970): 289–99.

39   J. F. Fulton, "Notes on a Recent Trip to England," *American Oxonian*, April 1941, pp. 72–92. These notes from a diary kept during a trip to England made on behalf of the U.S. National Research Council vividly portray the conditions in Oxford in October and November of 1940.

40   P. W. Clutterbuck, R. Lovell, and H. Raistrick, "Studies in the Biochemistry

of Microorganisms. No. 26. The Formation from Glucose by Members of the *Penicillin chrysogenum* Series of a Pigment, an Alkali-Soluble Protein, and Penicillin—The Antibacterial Substance of Fleming," *Biochem. J.* 26 (1932): 1907–18.

41 S. R. Craddock, Laboratory notebooks, unpublished. The British Library, Department of Manuscripts. Collection No. 56224 (Fleming), Vol. LI, 1924–31.

42 R. Hare, *The Birth of Penicillin* (Oxford: George Allen and Unwin, 1970), pp. 103–04, 169.

43 H. W. Florey, Personal communication to Dr. Gladys L. Hobby.

44 "From Our Own Correspondent—New York, March 13: Penicillin Research —'Momentous Result" from £320 Grant." *Times* (London), March 14, 1944.

45 J. Toland, *Adolf Hitler*, 2 vols. (New York: Doubleday & Company, 1976), pp. 714–40; 772–98; 794–96.

Chapter 5

1 J. Goldblum, archivist, National Research Council, Washington, D.C., personal communication to Dr. Gladys Hobby, August 1983.

2 C. Thom, "Mycology Presents Penicillin," *Mycologia* 37, no. 4 (August 1945): 460–75.

3 G. J. Dohrmann, "Fulton & Penicillin," *Surgical Neurology* 3, no. 5 (May 1975): 277–80.

4 P. Wells, "Some Aspects of the Early History of Penicillin in the United States," *J. Wash. Acad. Sci.* 65, no. 3 (1975): 96–101.

5 Northern Regional Research Center, Penicillin Chronology: Science & Education Administration, U.S. Department of Agriculture, compiled by Dean H. Mayberry, NC-173-80, U.S. National Archives, Washington, D.C. (1981).

6 R. D. Coghill, "The Development of Penicillin Strains," in *The History of Penicillin Production*, Chemical Engineering Progress Symposium, American Institute of Chemical Engineers, 66, no. 100 (1970): 1–21. Also, personal communication, Robert Coghill to Gladys Hobby, 1983.

7 N. G. Heatley, Letter to Dr. John F. Fulton, dated March 28, 1942, in *Personal Files of J. F. Fulton, Extracts of Correspondence* (with Sir Howard Florey, Lady Ethel Florey, Norman Heatley), Yale University Library.

8 Introduction to the Preliminary Inventory, NC-138 (TRG 227), National Archives, United States of America, Washington, D.C.

"The Office of Scientific Research and Development (OSRD) was created within the Office for Emergency Management by Executive Order 8807 of June 28, 1941, to assure adequate provision for research on scientific and medical problems relating to the national defense. The office was terminated on December 31, 1947, and its remaining functions were transferred for purposes of liquidation to the National Military Establishment by Executive Order 9913 of December 26, 1947. During the war period the Office served as a center for the mobilization of the scientific personnel and resources of the

Nation and it cooperated in planning, aiding, and supplementing, where necessary, the experimental and other research activities carried on by the armed services and other Federal agencies. To this end, it was authorized to enter into contracts and agreements with individuals, educational and scientific institutions (including the National Academy of Sciences and the National Research Council), industrial organizations, and other agencies for studies, experimental investigations, and reports. It was also given responsibility for similar contracts entered into, before its establishment, by the National Defense Research Committee (NDRC) and the Health and Medical Committee, which were created by order of the Council of National Defense on June 27 and September 19, 1940, respectively. It was responsible for contracts entered into by the Federal Security Administrator in his capacity as co-ordinator of health, medical, and related activities, as authorized by the Council of National Defense. The Office [of Scientific Research and Development] consisted of six major divisions: (1) the immediate Office of the Director, including an Advisory Council, charged with overall policy; (2) the Administrative Office under the Executive Secretary, charged with activities relating to budget and finance, personnel, the preparation of contracts, property control, and records; (3) the Liaison Office, responsible for scientific liaison with the Allied Powers, primarily Great Britain; (4) the National Defense Research Committee, charged with research and the development of new weapons and devices; (5) the Committee on Medical Research (CMR), charged with research in military medicine; and (6) the Office of Field Service, created by an administrative order on October 15, 1943, and charged with operational research, field engineering, the organization of laboratories established on military installations, and the analysis of field problems. Dr. Vannevar Bush, [President of the Carnegie Institute of Washington, D.C. and] Chairman of the NDRC since 1940, was appointed Director of the Office [of Scientific Research and Development].... Dr. J. B. Conant was named Chairman of the NDRC...and Dr. A. N. Richards was named Chairman of the CMR.... The primary function of the Administrative Office was to provide a supporting mechanism for the actions of the scientists of the National Defense Research Committee and the Committee on Medical Research.... The National Defense Research Committee (NDRC) was composed of eight members: The President of the National Academy of Sciences, the Commissioner of Patents, one representative each from the Army and the Navy and Drs. Karl T. Compton, James Bryant Conant, Richard Chase Tolman, and Vannevar Bush.... The Committee on Medical Research (CMR)...consisted of four members appointed by the President and three others designated, respectively, by the Secretary of War, the Secretary of the Navy, and the Administrator of the Federal Security Agency. Alfred N. Richards was Chairman; and the other civilian members were Lewis H. Weed, Alphonse R. Dochez, and A. Baird Hastings. Dr. Weed was named Vice Chairman and Irvin Stewart, Executive Secretary. After the cessation of hostilities, the Committee met infrequently until January 20, 1947, when it adjourned.... Research in the field of penicillin was encouraged by making some funds available to the

Department of Agriculture Laboratory in Peoria, Ill., in the summer of 1941 and thereafter. As penicillin became available, the CMR assumed responsibility for a clinical evaluation of the drug through an NRC committee. A program was also undertaken to synthesize penicillin. In October 1943 Dr. Bush appointed a Penicillin Committee, under Dr. Hans T. Clarke, to survey the field; and, as a result, contracts were let with various firms and colleges to study the chemical structure of penicillin."

9   Scientific personnel of the Committee on Medical Research of the OSRD: from *Advances in Military Medicine* made by American Investigators working under the Sponsorship of the CMR, ed. E. C. Andrus, D. W. Bronk, G. A. Cardin, Jr., C. S. Keefer, J. S. Lockwood, J. T. Wearn, and M. C. Winternitz, vol. 1 (Boston: Little, Brown & Co., 1948), pp. xi–xvii.

10 C. S. Keefer, "Penicillin: A Wartime Achievement," in *Science in World War II (Part Nine)—Office of Scientific Research and Development. Advances in Military Medicine* (Boston: Little, Brown & Co., 1948). In contrast to the OSRD which actually had authority to provide money for research conducted under contracts with private institutions and universities, the National Research Council in 1943 was only a quasi-governmental organization. It was an outgrowth of the National Academy of Sciences, which had been chartered by Abraham Lincoln. The National Research Council was founded during World War I to provide advisory scientific committees to government agencies. By 1943 there were a number of NRC drug and medical practice committees advising the military and one advising the War Production Board. The NRC Committee on Chemotherapy was chosen by the Office of Scientific Research and Development to handle all civilian use of penicillin and to correlate the reports resulting from such use. During World War II, Dr. Chester Keefer served as chairman of the NRC Committee on Chemotherapy. The evaluation of the therapeutic efficacy of penicillin that was carried out in civilian patients was conducted under the auspices of this committee. (See Food, Drug and Cosmetics Reports, August 28, 1943, U.S. National Archives, Record Group No. 277.)

11 I. Stewart, *Organizing Scientific Research for War—The Administrative History of the Office of Scientific Research and Development* (Boston: Little, Brown & Co., 1948).

12 C. Thom, "Molds, Mutants and Monographers," *Mycologia* 44, no. 1 (January–February 1952): 61–85.

13 K. B. Raper, "A Decade of Antibiotics in America," *Mycologia* 44, no. 1 (January–February 1952): 1–60.

14 K. B. Raper, "The Penicillin Saga Remembered," *ASM News* 44, no. 12 (1978): 645–53.

15 K. B. Raper, "Penicillin," *Yearbook of Agriculture* (1943–47): 699–710.

16 S. Mines, *Pfizer—an Informal History.* (New York: Pfizer, Inc., 1978), pp. 73–75.

17 M. Demerec, "Production of Penicillin," U.S. Patent Office Application No. 2,445,748, patented July 27, 1948 (Application filed September 16, 1946, Serial No. 697,380).

18 E. P. Abraham and E. Chain, "An Enzyme from Bacteria Able to Destroy Penicillin," *Nature* 146 (December 28, 1940): 837.

19 A. N. Richards, "Production of Penicillin in the United States (1941–1946)," *Nature* 201, no. 4918 (February 1, 1964): 441–45. Minutes of the first Conference on Penicillin, 1530 P Street, N.W., Washington D.C., October 8, 1941, read as follows: "Present: Dr. Bush, Dr. Richards, Dr. Weed, Dr. W. M. Clark, Dr. Thom, Dr. R. T. Major (Merck & Co.), Dr. George A. Harrop (E. R. Squibb & Sons), Mr. Kane (Pfizer & Co.), Dr. SubbaRow (Lederle & Co.).

"The purpose of the OSRD and CMR were explained by Dr. Bush, Dr. Richards and Dr. Weed. Dr. Thom described the part which the Peoria Laboratory is playing and the cooperation which it is prepared to give—study of many strains of penicillin in the hope of finding better ones than Fleming's. Dr. Major expressed the intention of his company to proceed to the point of 10,000 liters of culture a week. They are exceedingly willing to cooperate in every way.

"Harrop expressed the view that the Squibb Company was already far enough along to supply what was needed for clinical testing, which he thinks is the first requisite. He did not indicate any great willingness to share information or cooperate.

"Mr. Kane explained that they have been interested only a couple of weeks and that was through conversations with Dr. Dawson. He expressed the belief that his company would cooperate in any way which was desirable.

"Dr. SubbaRow explained that the Lederle Company was not interested in quantity production; they were afraid of it on the basis of possible contamination of their sera, etc., and that the small scale of laboratory experiments, which he is doing himself, were being done in a building one-half mile remote from the main building. His belief is that the best thing he can do at the present time is to continue to satisfy his own small scale chemical interest.

"Various questions were asked and answered concerning priorities and concerning patents. The patent sheets of both the NDRC and CMR contract forms were supplied to each. Dr. Bush left two questions with them (1) what each plans to do or is willing to do in this field, and (2) how far they are willing to go in collaboration. It was made clear that the Government had money to spend and they are to think of how the government can help them. The specific question was raised that, if anyone or all of the companies made sufficient product for both clinical and chemical testing, would they be willing to place enough for chemical study in the hands of certain university chemists.

"The party broke up with the thought that they will consult with the responsible heads of their companies, and will let me [i.e., Dr. Bush] know the answers to these questions or any others that have occurred to them with the thought that a further conference may be called. Dr. Bush emphasized the fact that we were not suggesting any commercial collusion at all; that we were only interested in the research aspects of the whole problem.

"Dr. Thom said that the group at the National Institutes of Health had a method of assay which he was sure would be turned over to people who needed it, and that he himself could dig up 12 or 20 different strains of penicillin [penicillium] which would be given out for study through Dr. [Chester] Emmons." Obviously no one of the participants at this meeting, not even Dr. Bush or Dr. Richards, recognized the magnitude of the problem that confronted them.

20 A. Fleming "On the Antibacterial Action of Cultures of Penicillium, with Special Reference to Their Use in the Isolation of B. Influenzae," *Brit. J. Exp. Path.* 10 (June 1929): 226–36.

21 P. W. Clutterbuck, R. Lovell, and H. Raistrick, "Studies in the Biochemistry of Microorganisms. No. 26. The Formation from Glucose by Members of the *Penicillium chrysogenum* Series of a Pigment, an Alkali-Soluble Protein, and Penicillin—The Antibacterial Substance of Fleming," *Biochem. J.* 26 (1932): 1907–18.

22 R. D. Reid, "Some Properties of a Bacterial-Inhibitory Substance Produced by a Mold," *J. Bact.* 29 (February 1935): 215–21.

23 E. Chain, H. W. Florey, A. D. Gardner, N. G. Heatley, M. A. Jennings, J. Orr-Ewing, and A. G. Sanders, "Penicillin as a Chemotherapeutic Agent," *Lancet* 2 (August 24, 1940): 226–28.

24 E. P. Abraham, E. Chain, C. M. Fletcher, H. W. Florey, A. D. Gardner, N. G. Heatley, and M. A. Jennings, "Further Observations on Penicillin," *Lancet* 2 (August 16, 1941): 177–88.

25 M. H. Dawson, G. L. Hobby, K. Meyer, and E. Chaffee, "Penicillin as a Chemotherapeutic Agent," *J. Clin. Invest.* 20, Abstr. (1941): 434.

26 G. L. Hobby, K. Meyer and E. Chaffee, "Activity of Penicillin in Vitro," *Proc. Soc. Expt. Biol. & Med.* 50 (1942): 277–80; "Observations on the Mechanism of Action of Penicillin," *Proc. Soc. Expt. Biol. & Med.* 50 (1942): 281–85.

27 C. McKee and G. Rake, "Activity of Penicillin against Strains of Pneumococci Resistant to Sulfonamide Drugs," *Proc. Soc. Exp. Biol. Med.* 51 (November 1942): 275–78.

28 C. McKee and C. L. Houck, "Induced Resistance to Penicillin of Cultures of Staphylococci, Pneumococci, and Streptococci," *Proc. Soc. Exp. Biol. Med.* 53 (May 1943): 33–34.

29 L. H. Schmidt and C. L. Sesler, "Development of Resistance to Penicillin by Pneumococci," *Proc. Soc. Exp. Biol. & Med.* 52 (April 1943): 353–57.

30 K. Meyer, G. L. Hobby, and M. H. Dawson, "The Chemotherapeutic Effect of Esters of Penicillin," *Proc. Soc. Expt. Biol. & Med.* 53 (1943): 100–04.

31 J. McIntosh and F. R. Selbie, "Zinc Peroxide, Proflavine and Pencillin in Experimental *Cl. welchii* Infections," *Lancet* 2 (December 26, 1942): 750–52.

32 J. McIntosh and F. R. Selbie, "Chemotherapeutic Drugs in Anaerobic Infections of Wounds," *Lancet* 1 (June 26, 1943): 793–95.

33 J. McIntosh and F. R. Selbie, "Combined Action of Antitoxin and Local Chemotherapy in *Cl. welchii* Infection in Mice," *Lancet* 2 (August 21, 1943): 224–25.

34 H. M. Powell and W. A. Jamieson, "Penicillin Chemotherapy of Mice

Infected with *Staphylococcus Aureus*," *J. Indiana Med. Assoc.* 35 (July 1942): 361–62.

35 F. R. Heilman and W. E. Herrell, "Penicillin in the Treatment of Experimental Relapsing Fever," *Proc. Staff Meetings of the Mayo Clinic* 18 (December 1, 1943): 457–67.

36 H. Welch, V. L. Chandler, R. P. Davis, and C. Price, "Study of Seven Different Salts of Penicillin," *J. Infect. Dis.* 76 (January–February 1945): 52–54.

37 D. H. Heilman and W. E. Herrell, "Comparative Antibacterial Activity of Penicillin and Gramicidin: Tissue Culture Studies," *Proc. Staff Meetings Mayo Clinic* 17 (May 27, 1942): 1–7.

38 W. E. Herrell and D. H. Heilman, "Tissue Culture Studies on Cytotoxicity of Bacterial Agents. I. Effects of Gramicidin, Tyrocidine, and Penicillin on Cultures of Mammalian Lymph Nodes," *Amer. Journ. of the Med. Sciences* 205, no. 2 (February 1943): 157–62.

Chapter 6

1 Howard Florey was later criticized by many for what was seen as "giving" penicillin to the Americans. Many resented the fact that the American pharmaceutical industry applied for patents on various steps in the processes developed for penicillin production, and in many cases those patents eventually were issued. No patent application to my knowledge ever applied, however, to the product itself. Further, it was American ingenuity, U.S. dollars, and U.S. production acumen that led ultimately to penicillin's availability as an effective chemotherapeutic drug (see chapters 5, 8, and 9).

2 Anonymous statement by Medical Research Council on Supplies and Distribution of Penicillin, *British Medical Journal*, August 28, 1943, p. 274.

3 L. P. Garrod, "Penicillin: Its Properties and Powers as a Therapeutic Agent," British Medical Bulletin 2, no. 1 (1944): 2–4.

4 M. E. Florey and H. W. Florey, "General and Local Administration of Penicillin," *Lancet* 1 (March 27, 1943): 387–97. See also *Special Contributions, Penicillin 1929–1943*, British Medical Bulletin 2, no. 1 (1944): 23–24.

5 C. S. Keefer, "Report on Penicillin Dated January 1, 1943. A Running Summary of the First 100 Cases of Infection Treated with Penicillin," U.S. National Archives, RG 227, OSRD (CMR).

6 Alexander Fleming, "Streptococcal Meningitis Treated with Penicillin: Measurement of Bacteriostatic Power of Blood and Cerebrospinal Fluid," *Lancet* 2 (September 10, 1943): 434–38. See also: *Special Contributions, Penicillin 1929–1943*, British Medical Bulletin 2, no. 1 (1944): 24–25.

7 D. C. Bodenham, "Infected Burns and Surface Wounds: The Value of Penicillin," *Lancet* 2 (November 12, 1943): 725–28. See also: *Special Contributions, Penicillin 1929–1943*, British Medical Bulletin 2, no. 1 (1944): 26–27.

8 A. M. Clark, L. Colebrook, T. Gibson, M. L. Thomson, and A. Foster, "Penicillin and Propamidine in Burns: Elimination of Hemolytic Streptococci

and Staphylococci," *Lancet* 1 (May 15, 1943): 605–09. See also: *Special Contributions, Penicillin 1929–1943*, British Medical Bulletin 2, no. 1 (1944): 24.

9 R. J. V. Pulvertaft, "Bacteriology of War Wounds," *Lancet* 2 (July 3, 1943): 1–2.

10 R. J. V. Pulvertaft, "Local Therapy of War Wounds with Penicillin," *Lancet* 2 (September 18, 1943): 341–46.

11 L. P. Garrod, "The Treatment of War Wounds with Penicillin," *British Medical Journal* 2 (December 11, 1943): 755–56.

12 H. W. Florey, Personal communication to Dr. Walsh McDermott, Cornell University Medical College, New York.

13 "Penicillin in War Wounds: A Report from the Mediterranean," *Lancet* 2 (December 11, 1943): 742.

14 H. W. Florey and H. Cairns, *Investigation of War Wounds. Penicillin. A Preliminary Report to the War Office and the Medical Research Council on Investigations Concerning the Use of Penicillin in War Wounds*, British War Office, A.M.D.7, 2 parts, October 1943, 114 pp.

15 L. G. Matthews, *History of Pharmacy in Britain* (Edinburgh and London: E. S. Livingstone, 1962), pp. 331–32.

16 The first collaborative research organized on a wide basis within the pharmaceutical industry in England came from a proposal made in 1941 to a number of companies that coordinated research in selected fields might have advantages, especially in wartime when scientific manpower in industry was already reduced by national service requirements. Five companies agreed to join in the proposed scheme—Boots Pure Drug Co., Ltd., British Drug Houses, Ltd., Glaxo Laboratories, Ltd., May and Baker, Ltd., and the Wellcome Foundation, Ltd. Together they formed the Therapeutic Research Corporation of Great Britain. It was agreed that the pooling of research would relate only to specified subjects. The participating companies would remain free to continue their individual research in other areas and would maintain separate existences as independent companies. A list of long-term and short-term projects for collaborative research was drawn up. By the time the program was organized, however, the production of penicillin became of such great importance that the members of the corporation, with some other companies, joined with university professors and chemists, and with members of a General Penicillin Committee set up the Ministry of Supply to investigate the nature of this new substance and ways for its production. Penicillin thus took first place in the corporation's activities. Other research was put aside, and in the months that followed, major contributions were made by member companies of the corporation to penicillin production and the study of its chemical nature.

17 H. W. Florey, E. B. Chain, N. G. Heatley, et al., *Antibiotics—A Survey of Penicillin, Streptomycin, and Other Antimicrobial Substances from Fungi, Actinomycetes, Bacteria and Plants*, 2 vols. (London: Oxford University Press, 1949), pp. 657–64.

18 C. Lyons, F. B. Queen, H. Hollenberg, J. S. Sweeney, and A. J. Ingram, *Penicillin Therapy for Septic Compound Fractures in a Military Hospital. A Report of an Investigative Unit of the Committee on Medical Research at the*

Bushnell General Hospital, U.S. National Archives, Record Group No. 227, Contractor's Technical Reports, OEM cmr-275, L4.

19 C. Lyons, *Penicillin Therapy of Surgical Infections in the U.S. Army*. Reprinted with additions from the *Journal of the American Medical Association* 123 (December 19, 1943): 1007–18.

Chapter 7

1 E. Chain, H. W. Florey, A. D. Gardner, and N. G. Heatley, "Penicillin as a Chemotherapeutic Agent," *Lancet* 2 (August 24, 1940): 226–28.

2 E. P. Abraham, E. Chain, C. M. Fletcher, H. W. Florey, A. D. Gardner, N. G. Heatley, and M. A. Jennings, "Further Observations on Penicillin," *Lancet* 2 (August 16, 1941): 177–88.

3 ICI (Dyestuffs), Limited, Research Department, "Report on a Meeting with Professor Florey and Dr. Heatley at Biochemical Laboratories at Oxford on 2nd June 1941," Unpublished report, courtesy of Imperial Chemical Industries.

4 N. G. Heatley, Personal communication to Gladys Hobby, 1982.

5 W. H. Helfand, H. B. Woodruff, K. M. H. Coleman, and D. L. Cowen, "Wartime Industrial Development of Penicillin in the United States," in *The History of Antibiotics*, ed. J. Parascandola, a symposium, sponsored by the Divisions of Chemistry and Medicinal Chemistry, American Chemical Society, April 5, 1979 (Madison, Wis.: American Institute of the History of Pharmacy, 1980), pp. 31–56.

6 A. D. Ainley, ICI (Dyestuffs), Limited. Technical Report. *Penicillin: Production at Trafford Park, 1942*. Interim. Research Department, dated October 27, 1942.

7 W. R. Boon, A. S. Gibson, C. M. Scott, and H. J. Thurlow, Imperial Chemical Industries, Ltd. Dyestuffs Division: *Joint Technical Report on Penicillin: Design, Erection, and Start-Up of the Initial Plant at Trafford Park Works, January to October, 1942*. Biological Department, Trafford Park Works, and Research Department Library File No. H.7051.

8 "The Production of Penicillin—Part I, A Short Account of Imperial Chemical Industries Plant," *Industrial Chemist*, November 1944, pp. 592–99.

9 H. Gudgeon and S. G. Terjesen, Imperial Chemical Industries, Ltd. (Dyestuffs Division). Technical Report. *Penicillin: Deep Culture Project*. Research Department, dated October 15, 1947. Library File No. R.7.1793.

10 C. T. Calam and G. C. M. Harris, Imperial Chemical Industries, Ltd., Biological Department (Penicillin Research), *Penicillin: Production by P. chrysogenum in Submerged Culture: Summary Report on Practical Aspects*. Report No. B.4, dated August 31, 1949.

11 L. G. Matthews, *History of Pharmacy in Britain* (Edinburgh and London: E. & S. Livingstone, Ltd., 1962).

12 B. Emery, Personal communications to Dr. Gladys L. Hobby, dated February 6, 1980, and March 20, 1980, based on a personal review by Bryan Emery of Glaxo Laboratories' World War II records.

13 P. W. Clutterbuck, R. Lovell, and H. Raistrick, "Studies in the Biochemistry of Microorganisms. The Formation from Glucose by Members of the Penicillium Chrysogenum Series of a Pigment, an Alkali-Soluble Protein, and Penicillin—The Antibacterial Substance of Fleming," *Biochemical J.* 26 (1932): 1907–18.

14 C. E. Coulthard, R. Michaelis, W. F. Short, G. Sykes, G. E. H. Skrimshire, A. F. B. Standfast, J. H. Birkinshaw, and H. Raistrick, "Notatin: An Antibacterial Glucose-Aerodehydrogenase from *Penicillin notatum* Westling," *Nature* 150 (1942): 634–35. Notatin, also known as Penatin Penicillin A, or Penicillin B, is a flavoprotein. Antibacterially active only in the presence of glucose, it is produced by some strains of *Penicillium notatum*.

15 H. T. Clarke, J. R. Johnson, and Sir Robert Robinson, eds., *The Chemistry of Penicillin* (Princeton, N.J.: Princeton University Press, 1949), pp. 99–105. Boon and his associates at the laboratories of ICI first used the distribution type of chromatogram described by Martin and Singe (*Biochem. J.* 38 [1944]: 224–32) for separation of penicillins. By this technique they isolated 2-pentenyl penicillin, known also as penicillin I (in Great Britain) or penicillin F (in the United States), and benzyl penicillin, commonly known as penicillin II or penicillin G. Penicillins F and G are two of the naturally occurring penicillins produced by various strains of penicillia.

16 L. G. Matthews, *History of Pharmacy in Britain* (Edinburgh & London: E. & S. Livingstone, Ltd., 1962), pp. 330–33.

17 National Research Development Corporation, *Evidence Offered to Review the Functioning of Financial Institutions. (The Wilson Committee).* (London: NDRC, 1977), pp. 3–5.

18 In the United States, filing of patent applications requires divulgence of information. Prior to World War II, some commercial firms therefore relied on "total secrecy" in lieu of patent coverage to protect their products. Certain of these corporations ultimately became involved in penicillin production or in other wartime activities. They shed their reliance on secrecy only slowly, for other products were to be considered. All this means that a great deal may have transpired in the early days of penicillin research that is not revealed in files on patents applied for or issued. Nonetheless, many applications for patents were filed—not on penicillin, but on methods for its production—and eventually many of these were issued, although not before 1947 to 1949. Of prime concern to the British were the following patents: (1) Dr. A. J. Moyer on the use of corn steep liquor in culture media used for penicillin production by surface and by submerged growth of appropriate strains of *Penicillia* (R. D. Coghill and A. J. Moyer, Peoria, Ill., "Method for Production of Increased Yields of Penicillin," Patented July 15, 1947—No. 2,423,873 [filed June 17, 1944]. Patent rights assigned to the United States of America, as represented by the Secretary of Agriculture. Also A. J. Moyer, Peoria, Ill., "Method for Production of Penicillin," Patented May 25, 1948— No. 2,442,141; also June 22, 1946—No. 2,443,989; and July 12, 1949—No. 2,476,107 [all filed May 11, 1945]. Patent rights assigned to the United States of America, as represented by the Secretary of Agriculture); (2) Dr. Milislav

Demerec on the mutation of penicillin-producing strains of *Penicillia* toward increased production of the substance (M. Demerec, Cold Spring Harbor, N.Y., "Production of Penicillin," Patented July 27, 1948—No. 2,445,748 [filed September 16, 1946]. Patent rights assigned to the United States of America, as represented by the Administrator, Civilian Production Administration); and (3) Dr. Otto K. Behrens of Eli Lilly and Company in Indianapolis, Ind., touching on virtually all phases of penicillin production technology but particularly on the fermentation of oxy-penicillins, a group of chemically related compounds which, under suitable conditions of growth, are produced as metabolic products by penicillin-producing molds, and on the use of phenylacetyl compounds as precursors to increase penicillin yields in fermentation liquors (O. K. Behrens, Indianapolis, Ind., "Process and Culture Media for Producing Penicillin," Patented April 27, 1948—No. 2,440,355 [filed April 21, 1945]. Patent rights assigned to Eli Lilly & Company, Inc. Also O. K. Behrens, J. W. Morse, R. E. Jones, and G. F. Soper, Indianapolis, Ind., "Process and Culture Media for Producing New Penicillins," Patented August 16, 1949—Nos. 2,479,295; 2,479,296; and 2,479,297 [all filed March 8, 1946]. Patent rights assigned to Eli Lilly & Company, Inc.). The sodium salt of penicillin was crystallized in 1943 by MacPhillamy, Wintersteiner, and Alicino of E. R. Squibb & Sons (O. Wintersteiner and H. B. MacPhillamy, New Brunswick, N.J., "Method of Obtaining a Crystalline Sodium Penicillin," Patented February 15, 1949—No. 2,461,949 [filed June 13, 1944]. Patent rights assigned to E. R. Squibb & Sons, New York). By 1948 fermentation yields of penicillin as high as 3,000 Oxford units per milliliter were reported by some groups (J. W. Foster, Austin, Texas, "Processes of Fermentation," Patented January 11, 1949—No. 2,458,495 [filed July 2, 1948]. Patent rights assigned to Merck & Co., Inc.). No record has been found to indicate what royalty payments resulting from issuance of these patents may have cost the British pharmaceutical industry. The drug was discovered and first developed in England—it was, as mentioned previously, Britain's major wartime contribution to society. But the British pharmaceutical industry had acquired its manufacturing know-how from the United States' pharmaceutical industry. Each owed the other much. (In England, patents are issued on the basis of who files first, not who is first to conceive the "invention.")

19 Medical Research Council War Memorandum No. 12, *The Use of Penicillin in Treating War Wounds* (Published by HMSO, 1944). The Penicillin Clinical Trials Committee, appointed by the Medical Research Council, consisted of Prof. H. R. Dean, M.D., F.R.C.P., chairman; and Prof. Drs. J. H. Burn, A. N. Drury, S. (Sir Sheldon) Dudley, A. Fleming, H. W. Florey, A. M. H. Gray, P. Hartley, R. Vaughan Hudson, G. L. Keynes, J. R. Learmonth, L. T. Poole, C. M. Scott, J. W. Trevan, R. V. Christie, and L. P. Garrod.

20 J. G. Barnes, "Early Work on Penicillin in England with Special Reference to Bromley," Unpublished report to Mr. M. W. Roche, dated May 26, 1976, from the files of E. F. Royds, Pfizer, Ltd., Sandwich, England.

21 E. F. Royds, Memo addressed to Mr. J. O. Teeter, Pfizer International Subsidiaries, dated July 22, 1971.

22 Kemball, Bishop & Co., Ltd., report summarizing "Penicillin: Its Development in the U.K. and U.S. as Described in Material from the Files of Kemball, Bishop," dated November 9, 1972.

23 On November 8, 1943, John L. Smith of Chas. Pfizer & Co., wrote to Kemball, Bishop, "We are working night and day to get the out-of-use ice plant...equipped for commercial production.... We have been, and are presently using, the flask process, the pan process, and the deep tank process. This is being done to acquire first-hand information regarding the merits and deportment of all three methods. In our commercial plant we are concentrating on the deep tank process." On March 8, 1944, he wrote again: "We have been tremendously busy in erecting and getting into production our new commercial penicillin plant. Here we use submerged fermentation, which we have been conducting on a pilot plant scale for about one year.... About two weeks ago we were visited by Mr. Jephcott and Dr. Boon...[and] discussed freely our penicillin process...and showed them...our deep tank fermentation plant." On December 20, 1945, Smith wrote, "In my last letter to Mr. [J. E.] Whitehall of Kemball, Bishop, I stated that in my opinion the deep tank Penicillin fermentation has important advantages over the bottle or tray processes...evidence is accumulating...[and] a culture has been isolated by one of the western universities which shows indications of giving a potency of 500 to 1000 units [per milliliter] in fermentation broth in deep tanks. I do not know what it would produce in bottles or trays because we have abandoned all work in this direction. I thought this information might be of interest, particularly at a time when you have expansion of the tray process under consideration." On April 23, 1946, the assistant controller of Penicillin Production Control, [British] Ministry of Supply wrote to Messrs. Kemball, Bishop & Co., Ltd., "In view of the greatly increased output of penicillin now being obtained from and the further increases in production to be expected from deep culture plants at Speke and Barnard Castle, following the introduction of the new deep culture strain Q.176 and the corresponding reduction in penicillin costs at these plants...the Ministry will not require the output from your plant after 1st July next."

24 S. Mines, *Pfizer—An Informal History* (New York: Pfizer Inc., 1978), pp. 42–44, 69–70.

25 L. M. Miall and V. J. Ward, Personal communications to Dr. Gladys L. Hobby, 1979 to 1984.

26 W. R. Boon and W. B. Hawes, Imperial Chemical Industries, Ltd. (Dyestuffs Division) and the Wellcome Foundation, Ltd. Technical Report. *Penicillin: Report of an Investigation into Research and Production in the United States of America, March–April 1943, Undertaken on Behalf of the Ministry of Supply.* (Unpublished report, Library File No. 5.895, dated November 25, 1943, courtesy of Imperial Chemical Industries and Dr. Boon.)

27 W. R. Boon and H. J. Thurlow, Imperial Chemical Industries, Ltd. (Dyestuffs Division). Technical Report. *Penicillin—Deep Culture Project.* (Unpublished report, Library File No. R.D.1793, dated October 15, 1947, courtesy of ICI.)

28 W. R. Boon and H. Jephcott, Imperial Chemical Industries, Ltd. (Dyestuffs

Division) and the Glaxo Laboratories. Technical Report. *Penicillin: Survey of Production Methods in United States of America and Canada, February/March 1944.* (Unpublished report dated May 31, 1945, courtesy of ICI and Dr. Boon.)

29 J. E. McKeen, "The Production of Penicillin," *Trans. Amer. Inst. of Chem. Engineers* 40, no. 6 (December 25, 1944): 747–58.

30 John Hastings, Personal communication, letter and report to Dr. G. L. Hobby, on the Speke plant, dated March 15, 1979.

31 J. C. Woodruff, Office memo, #40, 116—Penicillin, to Dr. Major, Merck & Co., Rahway, N.J., dated March 25, 1943.

32 Therapeutic Substances. Provisional & Statutory Rules & Orders 1944, No. 922, dated August 2, 1944, made by the Joint Committee constituted under Section 4(1) of the Therapeutic Substances Act, 1925 (15 & 16 Geo.5.C.6.) (published by HMSO, 1944).

33 A. A. Miles, Personal communication to Dr. Miall, Kemball, Bishop & Co., dated March 15, 1946, with statement on the history and establishment of the British National Standard for Penicillin.

34 Sir Henry Dale, Note to the Members of the General Penicillin Committee on New British Projects for Production of Penicillin by Submerged Culture, dated August 12, 1944. From the files of Messrs. Kemball, Bishop (courtesy of Mr. L. M. Miall, Pfizer, Ltd., Sandwich, England): "In August 1944, it was estimated that Distillers Co., Ltd. would produce 20,000 million units of penicillin per week at Speke (near Liverpool), Messrs. Boots, Ltd. would soon be able to produce 5–10,000 million units per week at Nottingham, and Messrs. Glaxo, Ltd. would produce at least 1500 million units at Barnard Castle in County Durham. In the first nine months of 1946, 1250 billion units were actually produced at the Speke plant alone."

35 F. Warburton, "Penicillin Sub-Committee on Additional Production," Interim Report—February 22 to April 18, 1944. From the Files of Messrs. Kemball, Bishop, Ltd. (courtesy of Mr. L. M. Miall, Pfizer, Ltd., Sandwich, England).

36 H. T. Clarke, J. R. Johnson, and Sir Robert Robinson, eds., *The Chemistry of Penicillin* (Princeton, N.J.: Princeton University Press, 1949), Preface, pp. v–vii: "As in the United States, a number of interrelated committees were needed to accomplish the goals of the penicillin program in England. The Therapeutic Research Corporation of Great Britain set up a Penicillin Sub-Committee of their Research Panel to deal with the production and chemistry of penicillin. The progress reports, rendered to this committee as occasion arose, were known as the 'PEN' reports.... In general, the chemical information contained in those reports was not published in the scientific press, but was privately communicated to recognized workers in the field.

"In October 1942, the Ministry of Supply set up a General Penicillin Committee. It was then decided that the existing Penicillin Sub-Committee should be enlarged to include other interested workers, among whom were Imperial Chemical (Pharmaceuticals) Ltd., and should continue its task of coordinating the work on production and purification of penicillin to the stage fit for clinical use, reporting to the General Penicillin Committee. The name of the Sub-

Committee was changed at this time to the Penicillin Producer's Conference or, as it was briefly called, the Penicillin Conference. At the same time it was arranged that the unofficial Conference of Chemists, which had been set up under the chairmanship of the late Dr. F. L. Pyman, should continue its function of handling information on the chemistry and structure of penicillin; the reports of this Chemists' Conference were made available to the Chairman of the General Penicillin Committee of the Ministry of Supply.

"By this time, experiments relating to the production of penicillin had been instituted in the laboratories of certain American pharmaceutical manufacturers (Merck & Co., Inc., E. R. Squibb & Sons, and subsequently Chas. Pfizer & Co., Inc.). The information secured by these firms was communicated to the Committee on Medical Research and transmitted by it, via the Medical Research Council, to the Therapeutic Research Corporation, Imperial Chemical Industries, and various academic groups in Britain.

"In October 1943, when many of the structural features of the penicillin molecule were becoming clear, the problem of the synthesis became a pressing one, and to handle this aspect the Medical Research Council set up a Committee for Penicillin Synthesis 'to initiate, coordinate and make investigations on the synthesis of penicillin and analogues.' The confidential reports which were issued and exchanged were known as the 'CPS' reports.

"At the same time, the Committee on Medical Research of the Office of Scientific Research and Development in Washington, which had already undertaken the coordination of chemical work on penicillin in the United States, came to an agreement with the Medical Research Council for an exchange of information on anything which had a bearing on the problem of the synthesis of penicillin. A limited group...of industrial and academic research organizations engaged, under contract with the Office of Scientific Research and Development, to collaborate in studies 'in connection with (i) the chemical structure of penicillin and (ii) the synthesis of penicillin or a therapeutic equivalent.'"

Chapter 8

1   "Penicillin Output Remains Far Below Desired Goals," *Journal of Commerce*, August 5, 1943.
2   A. N. Richards, "Production of Penicillin in the United States (1941–1946)," *Nature* 201, no. 4918 (February 1, 1964): 441–45.
3   Introduction to the Preliminary Inventory, NC-138 (TRG 227), U.S. National Archives, Washington, D.C.
4   C. S. Keefer, "Penicillin: A Wartime Achievement," in *Science in World War II: Advances in Military Medicine* (Boston: Little, Brown & Co., 1948), pt. 9, "Penicillin."
5   I. Stewart, *Organizing Scientific Research for War—The Administrative History of the Office of Scientific Research and Development* (Boston: Little, Brown & Co., 1948).

6 James Phinney Baxter, 3d, *Scientists against Time* (Boston: Little, Brown & Co., 1947), pp. 337–59.

7 "Statement by the [British] Medical Research Council on Supplies and Distribution of Penicillin," *British Medical Journal* (August 28, 1943): 274.

8 C. S. Keefer, "Report on Penicillin, Dated January 1, 1943. A Running Summary of the First 100 Cases of Infection Treated with Penicillin," U.S. National Archives, RG227, OSRD (CMR).

9 C. S. Keefer, F. G. Blake, E. K. Marshall, Jr., J. S. Lockwood, and W. B. Wood, Jr., "Penicillin in the Treatment of Infections: A Report of 500 Cases," *J. Amer. Med. Assoc.* 122 (August 28, 1943): 1217–.

10 F. J. Stock, "Production and Distribution of Penicillin," *Amer. Pharmac. Assn. Journ.* (Practical Pharmacy Edition) 6 (1945): 110–14.

11 War Production Board, Advance Release, Office of War Information, WPB-7477, March 8, 1945.

12 D. C. Anderson and C. S. Keefer, *The Therapeutic Value of Penicillin—A Study of 10,000 Cases* (Ann Arbor, Mich.: J. W. Edwards, 1948), pp. v–viii.

13 M. Tager, "John F. Fulton, Coccidiomycosis, and Penicillin," *Yale Journal of Biology and Medicine* 19, no. 4 (September 1976): 391–98.

14 J. F. Fulton, *Personal Diary*, vol. XVII, 1942, Historical Library, Yale University Medical Library, New Haven, Conn.

15 G. J. Dohrmann, "Fulton and Penicillin," *Surgical Neurology* 3, no. 5 (May 1975): 277–80.

16 A. Fleming "On the Antibacterial Action of Cultures of Penicillium, with Special Reference to Their Use in the Isolation of *B. influenzae*," *Brit. J. Exp. Path.* 10 (June 1929): 226–36.

17 R. L. Cecil, ed., *A Textbook of Medicine by American Authors* (Philadelphia and London: W. B. Saunders Co., 1943).

18 W. McDermott and D. E. Rogers, "Social Ramifications of Control of Microbial Disease," *Johns Hopkins Medical Journal* 151 (1982): 302–12.

19 Gladys L. Hobby, Unpublished materials, personal laboratory records.

20 M. H. Dawson and G. L. Hobby, "The Clinical Use of Penicillin—Observations in One Hundred Cases," *J. Amer. Med. Assoc.* 124 (March 4, 1944): 611–22.

"Among 100 patients treated with penicillin, the first of three with pneumococcal endocarditis came under observation in March 1942. Treatment with Pfizer's penicillin lot No. 792 was initiated on the 26th of the month. A man aged 53 was apparently recovering uneventfully from a lobar pneumonia (type 7) when he developed a septic temperature. Sulfonamide therapy in adequate dosage failed to improve the situation and a blood culture revealed 650 colonies of pneumococcus (type 7) per cubic centimeter [i.e., per milliliter]. The patient was given approximately 10,000 units of penicillin every three hours, intravenously. Within twenty-four hours an astonishing improvement in the clinical condition was observed. There was a change from a comatose state to one of mental alertness, the temperature returned to normal, and a blood culture taken at the end of the first day was negative. Improvement continued for a further period of forty-eight hours, but the supply of penicillin available was so limited that it was necessary to reduce the dose to 5,000 units every three hours. At the end of 72 hours of treatment, there was a recurrence of fever, and a blood culture showed 20 colonies per cubic centimeter. The dose of penicillin was again increased to 10,000 units every three hours and this was followed by a satisfactory improvement in the clinical condition. A negative blood culture was obtained a second time. After a further period of time, however, it became obvious that the infection could not be controlled with the amount of penicillin available, and therapy was discontinued. In the second case of acute pneumococcic endocarditis, similar results were obtained. After the administration of 30,000 units of penicillin by infusion in the first two hours, followed by 10,000 units every four hours for three doses, a negative blood culture was obtained. A total of 175,000 units (Merck's penicillin lot No. 41922) was given in the first three days and there was temporary improvement in the patient's condition. Again it became apparent...that the infection could not be controlled with the quantity of penicillin available and therapy was discontinued."

21 W. S. Tillett, J. E. McCormack, and M. J. Cambier, "The Treatment of Lobar Pneumonia with Penicillin," *J. Clin. Invest.* 24 (1945): 589–94.

22 W. S. Tillett, J. E. McCormack, and M. J. Cambier, "The Use of Penicillin in the Local Treatment of Pneumococcal Empyema," *J. Clin. Invest.* 24 (1945): 595–610.

23 C. Lyons, *Penicillin Therapy of Surgical Infections in the U.S. Army*, U.S. National Archives, Record Group No. 227, Contractor's Technical Reports, OEM cmr-275, L4.

24 C. Lyons, F. B. Queen, H. Hollenberg, J. S. Sweeney, and A. J. Ingram, *Penicillin Therapy for Septic Compound Fractures in a Military Hospital. A Report of an Investigative Unit of the Committee on Medical Research at the Bushnell General Hospital*, U.S. National Archives, Record Group No. 227, Contractor's Technical Reports, OEM cmr-275, L4.

25 C. Lyons, *Penicillin Therapy of Surgical Infections in the U.S. Army*. Reprinted with additions from the *Journal of the American Medical Association* 123 (December 18, 1943): 1007–18.

26 P. B. Beeson, W. McDermott, and J. B. Wyngaarden, eds., *Cecil Textbook of Medicine*. 15th ed. (Philadelphia, London, and Toronto: W. B. Saunders Co., 1979).

27 J. E. Moore, "Venereal Diseases," in *Advances in Military Medicine*, Chapter III (Boston: Little, Brown & Company, 1948), pp. 46–64.

28 C. S. Keefer, "Penicillin," *Proc. of the Institute of Medicine of Chicago* 15, no. 17 (November 15, 1945): 9.

29 C. S. Keefer, "The Present Status of Penicillin in the Treatment of Infections," *Proc. the Amer. Philosoph. Soc.* 88, no. 3 (September 1944): 174–76.

30 W. H. D. Priest, "The Treatment of Gonorrhea with Penicillin," in H. W. Florey and H. Cairns: *Investigation of War Wounds. Penicillin. A Preliminary Report to the War Office and the Medical Research Council Concerning the Use of Penicillin in War Wounds*, British War Office, A.M.D.7, October 1943, sec. B, Pt. I, p. 27, and sec. 33, pt. II, pp. 105–12.

31 "Syphilis and Gonorrhea," *VD Fact Sheet, 1976*, Disease Trends, CDC. Ed. 33 (Washington D.C.: U.S. Department of Health, Education and Welfare, Public Health Service, 1941–76).

32 M. S. Siegel, S. E. Thompson, and P. L. Perine, "Penicillinase-Producing *Neisseria gonorrhoeae*," *Sex Transm. Dis.* 4 (1977): 32–33.

33 H. H. Handsfield, E. G. Sandström, J. S. Knapp, P. L. Perine, G. B. Whittington, D. E. Sayers, and K. K. Holmes, "Epidemiology of Penicillinase-Producing *Neisseria Gonorrhoeae* Infections," *New Eng. J. Med.* 306, no. 16 (April 22, 1982): 950–54.

34 Center for Disease Control, "Surveillance Summary on Penicillinase-Producing *Neisseria gonorrhoeae*—United States, Worldwide," *Morbidity and Mortality Weekly Report* 28, no. 8 (March 2, 1979): 85–87.

35 C. S. Keefer, "Penicillin: A Wartime Achievement," in *Science in World War II (Part Nine)—Office of Scientific Research and Development. Advances in Military Medicine* (Boston: Little, Brown and Company, 1948).

36 R. C. Arnold, "The Development of the Concept of Mass Treatment of Syphilis with Penicillin," Report to the World Forum on Syphilis, 1962.

37 C. S. Keefer, "Memorandum on the Use of Penicillin in Syphilis," to Dr. Alfred N. Richards, National Research Council (U.S.A.), dated October 27, 1943, U.S. National Archives, RDG 227 (CMR).

38 R. C. Williams, *The United States Public Health Service, 1798–1950* (Bethesda, Md.: Commissioned Officers Association of the U.S. Public Health Service, 1951), pp. 387–90.

39 James Phinney Baxter, 3d, *Scientists against Time*, Chapter XXII (Boston: Little, Brown & Company, 1947), pp. 337–59.

40 "Disease Trends, Syphilis and Gonorrhea, 1941–1976," *VD Fact Sheet 1976*, 33d ed., HEW Publication No. (CDC) 77–8195 (Washington, D.C.: U.S. Department of Health, Education and Welfare, Public Health Service, Center for Disease Control, 1976). In 1953, Dr. Leonard Heimoff (former venereal disease control officer for the western division of the Pacific theater of operations during World War II) remarked that when he had entered the Army in 1941, the venereal disease rate was about 40 per 1,000 persons per annum, and gonorrhea was by far the most frequent of the venereal diseases. A report from the navy indicated that there were about 30 cases of gonorrhea to one of the syphilis, and a report from the army in 1946 showed the ratio to be 6 cases of gonorrhea to 1 of syphilis. On the basis of 40 cases per 1,000 persons, an army of 8 million (as was planned for in 1943) would have had to deal with approximately 400,000 cases of venereal disease each year, by far the largest proportion of the cases being gonorrhea. The treatment of gonorrhea at the time was largely symptomatic, and the disease ran a course of about three months. That was prior to the advent of sulfanilamide. With the application of sulfanilamide, it was possible to obtain cures in five days. Difficulties arose, however, due to drug toxicity and the emergence of microbial drug resistance, and by late 1943, many military installations reported cure rates for gonorrhea as low as 30 percent. Fortunately, the effectiveness of penicillin in the treatment of gonorrhea was demonstrated at about this time (1943), and in

1953 Heimoff reported cure rates for penicillin-treated gonorrhea approximating 90 percent (see L. Heimoff, *New York State Journal of Medicine*, July 1, 1953, pp. 1564–72). The appearance of penicillinase-producing strains of gonococci came later and to some extent diminished the drug's effectiveness, thus forcing the use of higher dosages.

41 *Summary of Vital Statistics 1979: The City of New York* (New York: Department of Health, City of New York, Bureau of Health Statistics and Analysis, 1979).

42 F. W. Reynolds and J. E. Moore, "Syphilis, a Review of the Recent Literature," *Archives of Internal Medicine* 80 (November 1947): 655–90; 80 (December 1947): 799–840; 81 (January 1948): 85–108.

43 J. F. Mahoney, R. C. Arnold, and A. Harris, "Penicillin Treatment of Early Syphilis: A Preliminary Report," *Ven. Dis. Inform.* 24 (December 1943): 355–57.

44 J. F. Mahoney, R. C. Arnold, B. L. Sterner, A. Harris, and M. R. Zwally, "Penicillin Treatment of Early Syphilis, II," *J. Amer. Med. Assoc.* 126, no. 2 (September 9, 1944): 63–67.

45 J. E. Moore, J. F. Mahoney, W. Schwartz, T. Sternberg, and W. B. Wood, "The Treatment of Early Syphilis with Penicillin: A Preliminary Report of 1,418 Cases," *J. Amer. Med. Assoc.* 126, no. 2 (September 9, 1944): 67–73.

46 J. H. Stokes, T. H. Sternberg, W. H. Schwartz, J. F. Mahoney, J. E. Moore, and W. B. Wood, "The Action of Penicillin in Late Syphilis, Including Neurosyphilis, Benign Late Syphilis, and Late Congenital Syphilis: Preliminary Report," *J. Amer. Med. Assoc.* 126, no. 2 (September 9, 1944): 73–80.

47 G. Seger, "Status of the Volunteer Study of Penicillin in the Treatment of Early Syphilis," *Report (S-10747-C) of the Executive Assistant, Syphilis Study Section, National Institutes of Health (U.S.A.)*, dated September 24, 1947.

48 J. E. Moore, *Penicillin in Syphilis* (Springfield, Ill.: Charles C. Thomas, 1946).

49 G. L. M. McElligott, "Venereal Diseases," in A. Fleming, ed., *Penicillin—Its Practical Application* (London: Butterworth & Co., 1950), pp. 289–302.

50 R. C. Arnold, Letter to Dr. Gladys L. Hobby, dated December 4, 1978.

51 Steven M. Spencer, "The Great Penicillin Mystery," *Saturday Evening Post*, August 3, 1946, pp. 10, 52, 54–55.

52 M. P. Backus, J. F. Stauffer, and M. J. Johnson, "Penicillin Yields from New Mold Strains," *Amer. Chem. Soc. Journ.* 68 (1946): 152–53.

53 G. L. Hobby, T. F. Lenert, and B. Hyman, "The Effect of Impurities on the Chemotherapeutic Action of Crystalline Penicillin," *J. Bacteriol.* 54, no. 3 (September 1947): 305–23.

54 R. Tompsett, S. Schultz, and W. McDermott, "The Relation of Protein Binding to the Pharmacology and Antibacterial Activity of Penicillins X, G, Dihydro- F, and K," *J. Bacteriol.* 53 (1947): 581–95.

55 H. T. Clarke, J. R. Johnson, and Sir Robert Robinson, eds., *The Chemistry of Penicillin. Report on a Collaborative Investigation by American and British Chemists Under the Joint Sponsorship of the Office of Scientific Research and Development, Washington, D.C., and The Medical Research Council, London* (Princeton, N.J.: Princeton University Press, 1949).

292

56 M. H. Dawson and T. H. Hunter, "The Treatment of Subacute Bacterial Endocarditis with Penicillin—Results in Twenty Cases," *J. Amer. Med. Assoc.* 127 (January 20, 1945): 129–37.

57 G. L. Hobby, Personal records and recollections.

58 P. DeKruif, "Conquest of a Killer," *Reader's Digest,* March 1945, pp. 27–31.

59 L. Bickel, *Rise Up to Life* (London: Angus and Robertson, 1972).

60 J. D. Ratcliff, *Yellow Magic, the Story of Penicillin* (New York, Random House, 1945).

61 P. DeKruif, "Leo the Bold," in *Life among the Doctors* (New York: Harcourt, Brace & Co., 1949), pp. 210–51.

62 L. Loewe, P. Rosenblatt, H. J. Green, and M. Russell, "Combined Penicillin and Heparin Therapy of Subacute Bacterial Endocarditis," *J. Amer. Med. Assoc.* 124 (January 15, 1944): 144–49.

63 L. Loewe, P. Rosenblatt, M. Russel, and E. Alture-Werber, "The Superiority of the Continuous Intravenous Drip for the Maintenance of Effectual Serum Levels of Penicillin: Comparative Studies with Particular Reference to Fractional and Continuous Intramuscular Administration," *J. Lab and Clin. Med.* 30, no. 9 (September 1945): 730–35.

64 W. J. MacNeal, A. Blevins, and C. A. Poindexter, "Clinical Arrest of Endocarditis Lenta by Penicillin," *Amer. Heart Journ.* 28, no. 5 (November 1944): 669–79.

65 D. G. Anderson and C. S. Keefer, "The Treatment of Nonhemolytic Streptococcus Subacute Bacterial Endocarditis with Penicillin," *Medical Clinics of North America* (September 1945): 1129–52.

66 R. V. Christie, "Subacute Bacterial Endocarditis," in A. Fleming, ed., *Penicillin—Its Practical Application* (London: Butterworth & Co., Ltd., 1950), pp. 132–37.

67 Penicillin was not synthesized until 1959 (J. C. Sheehan and K. R. Henery-Logan, "The Total Synthesis of Penicillin V," *J. Amer. Chem. Soc.* 81 [1959]: 3089–94), and even today (1982), penicillin is produced by fermentation for commercial purposes.

68 During the next five- to ten-year period, the pharmacological action of penicillin was studied intensively (C. H. Rammelkamp and C. S. Keefer, "The Absorption, Excretion and Distribution of Penicillin," *J. Clin. Invest.* 22 [1943]: 425–37). Attention was directed to appropriate dosage regimens and time-dose relationships in the treatment of infectious disease (E. K. Marshall, Jr., "The Dosage Schedule of Penicillin in Bacterial Infections," *Bull. of the Johns Hopkins Hospital* 82, no. 3 [March 1948]: 403–07). The need for absorption-delaying agents became apparent, and Dr. Monroe Romansky— on D-Day, June 6, 1944—first obtained (in rabbits) evidence of the prolonged action of penicillin when administered in oil and beeswax (Personal communication from M. J. Romansky to G. L. Hobby, dated September 21, 1978). This led to the widespread use of penicillin in oil and beeswax (M. J. Romansky, "Penicillin. [1] Prolonged Action in Beeswax-Peanut Oil Mixture. [2] Single Injection Treatment of Gonorrhea," *U.S. Army Med. Bull.* 810 [October 1944]: 43–49). The use of this formulation, extremely irritating to

body tissues and painful as it was, continued until the introduction of procaine penicillin and other absorption-delaying forms of the drug (G. L. Hobby, E. Brown, and R. A. Patelski, "Biological Acitivity of Crystalline Procaine Penicillin *In Vitro* and *In Vivo*," *Proc. Soc. Expt. Biol. Med.* 67 [1948]: 6–14).

During this interim period, too, it was established that penicillin—despite its instability at the pH of gastric juice—could be administered effectively by the oral route, provided an adequately large dose of the drug is used. The earliest observations were made by Dr. R. L. Libby of Lederle Laboratories (R. L. Libby, "Oral Administration of Penicillin in Oil," *Science* 101 [1945]: 178). Later, Dr. Walsh McDermott and his associates established that it was necessary to administer approximately five times as much penicillin by the oral as by the intramuscular route in order to obtain comparable concentrations of penicillin in the blood. From McDermott's carefully designed studies, it was also established that a period of at least two hours should elapse between a meal and the ensuing dose of penicillin (W. McDermott, P. A. Bunn, M. Benoit, R. DuBois, and W. Haynes, "Oral Penicillin," *Science* 101 [1945]: 228. Also W. McDermott, P. A. Bunn, W. Benoit, R. DuBois, and M. E. Reynolds, "The Absorption of Orally Administered Penicillin," *Science* 101 [1946]: 359).

It was during this period, too, that interest first began to center on the emergence of acquired bacterial resistance to penicillin. Early observations in this regard were summarized by Lawrence P. Garrod in 1950 (L. P. Garrod, "Acquired Bacterial Resistance to Chemotherapeutic Agents," *Bull. Hyg.* 25, no. 6 [June 1950]: 539–54). As pointed out by Garrod, "The first clue to the nature of this bacterial resistance was provided by the work of Kirby...who [in 1944 and 1945] extracted from...such strains a potent penicillin inactivator." By 1946 the distinction between natural resistance and acquired resistance began to be recognized. In later years, the study of the mechanisms responsible for microbial resistance to penicillin has assumed great importance.

Chapter 9

1 War Production Board, Allocation Order M-338, dated July 9, 1943; War Production Board 6670, U.S. Government Printing Office, p. 1, Part 3281-Penicillin. From the U.S. National Archives, Records Group No. 227.

2 C. S. Keefer, "Penicillin: A Wartime Achievement," in *Science in World War II: Advances in Military Medicine*, Part 9, Ch. L11 (Boston: Little, Brown & Co., 1948), pp. 717–19.

3 J. P. Baxter, 3d, "Penicillin," in *Scientists Against Time* (Boston: Little, Brown & Co., 1947), ch. 22.

4 Albert Elder, "Penicillin," *Scientific Monthly* 58 (June 1944): 405–09. See also Economic Report on Antibiotics Manufacture, Federal Trade Commission Report (Washington, D.C.: U.S. Government Printing Office, June 1958), pp. 43, 334.

5 H. W. Florey, E. Chain, N. G. Heatley, M. A. Jennings, A. G. Sanders, E. P. Abraham, and M. E. Florey, *Antibiotics: A Survey of Penicillin, Streptomycin, and Other Antimicrobial Substances from Fungi, Actinomycetes, Bacteria and Plants* (London, New York, and Ontario: Oxford University Press, 1949), pp. 649–50.

6 J. D. Ratcliff, *Yellow Magic: The Story of Penicillin*, foreword by C. S. Keefer (New York: Random House, 1945), pp. ix–x.

7 F. Stock, Personal communication to Gladys Hobby, dated October 30, 1978: "I think you have a copy of *Yellow Magic* by Ratcliff. Before writing the book he talked to me and members of my staff. The introduction, written by Dr. Chester Keefer, is factual as to the companies. I supplied the names to Chester."

8 A. Fleming, "On the Antibacterial Action of Cultures of a Penicillium, with Special Reference to Their Use in the Isolation of *H. influenzae*," *Brit. J. Exp. Path.* 10, no. 3 (1929): 226–36.

9 P. W. Clutterbuck, R. Lovell, and H. Raistrick, "Studies in the Biochemistry of Microorganisms—No. 26—The Formation from Glucose by Members of the *Penicillium chrysogenum* Series of a Pigment, an Alcohol-Soluble Protein, and Penicillin—the Antibacterial Substance of Fleming," *Biochem. J.* 26 (1932): 1907–18.

10 Gladys L. Hobby, Personal laboratory notebooks and records.

11 R. D. Coghill to A. N. Richards, Letter dated December 27, 1941, U.S. National Archives, Record Group No. 227.

12 G. L. Hobby, "Early Investigations of Penicillin" (Columbia University and Chas. Pfizer & Co.). Prepared July 11, 1949, from personal records: "Early work of Dawson, Hobby and Meyer: First studies on fermentation of penicillin started on September 23, 1940. The culture used on this day had been received from Roger Reid of the Johns Hopkins Hospital. Culture had been received by Dr. Reid from Fleming in 1935. First series of flasks showed activity after 4 days and first sensitivity tests, using green streptococci, from cases of subacute bacterial endocarditis, were run on Sept. 30, 1940. First toxicity and protection tests were run on October 2, 1940, using active broth filtrates. First concentrates were parpared by Dr. Meyer on October 7, 1940. First injection in patients on January 11, 1941.

"Culture was received direct from Dr. Chain and receipt of it acknowledged October 28, 1940, with statement that work was started with Reid's strain of Fleming's culture. Chain's culture showed purplish red sporulation on receipt; on subculture it grew with grayish green sporulation; never produced any penicillin in our hands....

"*On June 6, 1941*—Mr. G. O. Cragwall of Chas. Pfizer & Co., Inc. wrote to Dr. Dawson stating the company's interest in the paper which was presented at the meeting of the American Society of Clinical Investigation.... This letter was acknowledged by Dr. Dawson on June 9, 1941.

"*On June 17, 1941*—Mr. Cragwall again wrote to Dr. Dawson—'I note that at the present time you are amply supplied with material for use in your investigation and do not believe that we can be of any service now. If at any time in the future you would find that our experience in and facilities for preparing large growths of molds and other organisms may be of service, we shall be very pleased to hear from you.' Insofar as I recall, or my records show, Chas. Pfizer & Co., Inc. was the only company that contacted Dr. Dawson following presentation of the paper on penicillin on May 5, 1941.

"*On July 31, 1941*—Dr. Dawson wrote to Mr. Cragwall stating 'Since my recent letter to you there has been significant progress in the research work on Penicillin, and there is now a possibility that we would be much interested in the production of this product on a large scale.... There is a real possibility that we will be interested in discussing the problem with you in the near future. As soon as further details are available I will let you know.'

"*On September 18, 1941*—Letter from Mr. Cragwall to Dr. Dawson stating that Mr. [J. H.] Kane, Mr. [J.] Davenport, and Mr. [G.] Cragwall will visit Dr. Dawson on September 24th.

"*On October 1, 1941*—Letter from Mr. Kane to Dr. Hobby—'Kindly deliver to bearer the media and tube of culture which we discussed on the telephone this morning.'

"*On October 1, 1941*—Laboratory notebook reads '1st lot to Pfizer—CLXII—1:320 titer—about 7–8 liters sent to Pfizer & Co. [for them] to test method of concentrating.'

"*October 3, 1941*—Laboratory notebook reads Pfizer concentrates made [by them] from CLXII (active in a dilution of 1:640) were received and tested by Hobby for activity.

"*October 2, 3, and 4th, 1941*—Letters received by Dr. Hobby from Mr. Kane indicating that samples were sent by Pfizer to Dr. Hobby for assay.

"From then on records show an exchange of penicillin liquor and samples between Chas. Pfizer & Co., and P&S. At that time it was difficult to interpret differences in titrations at the two laboratories; now it is apparent that the differences were due to the fact that Pfizer was using a strain of staphylococcus for titration and we were using a strain of hemolytic streptococci; these two strains differ quantitatively in their sensitivity to the penicillins.

"*October 3, 1941*—Cylinders, cultures, etc. received from Heatley by Dr. Hobby.

"*On November 25, 1941*—Letter from Dr. Hobby to Mr. Kane requesting that concentrates be forwarded to Dr. Meyer instead of crude fermentation liquors if possible.

"*November 28, 1941*—Letter from Mr. Kane to Dr. Hobby stating that it will not be possible to send concentrates or extracts in view of the fact that Dr. Walton Smith was using all that they could produce.

"*January 29, 1942*—Letter from Mr. Kane to Dr. Hobby discussing comparison of Pfizer penicillin and P&S penicillin with Oxford penicillin standard.

"*February 26, 1942*—Letter from Mr. [John L.] Smith to Dr. Dawson—Discontinues delivery of fermentation liquor and started delivery of '150–200 milligrams of penicillin concentrate weekly to be used for the clinical testing of cases which you may have at your disposal.' (This was with approval of Dr.

A. N. Richards.)

"*March 2, 1942*—First delivery of dry penicillin (500 mgs.). See letter from Mr. Smith to Dr. Dawson on March 5, 1942—Lot #696: Order No. 1045. Ampuls sealed in vacuo.

"*March 5, 1942*—First injection of Pfizer Penicillin Lot #696 to patients. Used almost daily thereafter.

"*April 4, 1942*—First report to Mr. Smith from Dr. Dawson on clinical results with Pfizer penicillin.

"*November 3, 1942*—All delivery of Pfizer penicillin to P&S discontinued."

13 M. H. Dawson to E. B. Chain, Letter dated October 28, 1940; in personal files of G. L. Hobby.

14 W. H. Helfand, H. B. Woodruff, K. M. Coleman, and D. L. Cowen, "Wartime Industrial Development of Penicillin in the United States," in *The History of Antibiotics—A Symposium*, ed. J. Parascandola (Madison, Wis.: American Institute of the History of Pharmacy, 1980).

15 S. Waksman, *Microbial Antagonisms and Antibiotics* (New York: Commonwealth Fund, 1945).

16 M. H. Dawson, G. L. Hobby, K. Meyer and E. Chaffee, "Penicillin as a Chemotherapeutic Agent," presented at the Annual Meeting of the American Society of Clinical Investigation, May 5, 1941.

17 G. O. Cragwall to M. H. Dawson, Letter dated June 6, 1941; in personal files of G. L. Hobby; also in historical records of Chas. Pfizer & Co., compiled by G. B. Stone, vol. III, for the period 1940–45.

18 R. D. Coghill to G. L. Hobby, Personal communication, 1983. Dr. Coghill became director of the Northern Regional Research Laboratory in 1938, soon after its authorization by Congress. Penicillin was the first major project at the Peoria laboratories.

19 R. D. Coghill, "The Development of Penicillin Strains," in *The History of Penicillin Production*, Chemical Engineering Progress Symposium Series, no. 100 (New York: American Institute of Chemical Engineers, 1970), 66: 14–21. Also R. D. Coghill to G. L. Hobby, Personal communication, 1983.

20 S. Mines, *Pfizer—An Informal History* (New York: Pfizer, 1970).

21 A. C. Finlay, Transcript of oral history of Pfizer, Inc., recorded August 18, 1970. Also see Pfizer historical records, and S. Mines, *Pfizer—An Informal History* (New York: Pfizer, Inc., 1970).

22 R. S. Sellers, Transcript of oral history of Pfizer, Inc. Also see unpublished Pfizer historical records, and S. Mines, *Pfizer—An Informal History* (New York: Pfizer, Inc., 1970).

23 A. N. Richards to G. H. A. Clowes, Letter dated December 20, 1941. U.S. National Archives, Record Group No. 227.

24 A. B. Hastings, Notes on a meeting held December 17, 1941, to discuss penicillin production. U.S. National Archives, Record Group No. 227.

25 J. Sheehan and R. N. Ross, "The Fire That Made Penicillin Famous," *Yankee*, November 1982, pp. 124–203.

26 Medical Research Council, News Letter No. 26, "Boston Fire Disaster and

Visit to Rochester," prepared by J. H. Burn, Liaison Office, Washington, D.C., December 17, 1942. U.S. National Archives, Record Group No. 227.

27 M. Tager, "John F. Fulton, Coccidiomycosis, and Penicillin," *Yale Journal of Biology and Medicine* 19, no. 4 (September 1976): 391–98.

28 J. F. Fulton, *Personal Diary*, vol. XVII, 1942. Historical Library, Yale University Medical Library, New Haven, Conn.

29 G. J. Dohrmann, "Fulton and Penicillin," *Surgical Neurology* 3, no. 5 (May 1975): 277–80.

30 R. D. Coghill to A. N. Richards, Letter dated December 27, 1941. U.S. National Archives, Record Group No. 227. "Chas. Pfizer & Co. is not working on the scale that Merck is, but they may be in a position to make a real contribution in the work. Their organism is not producing the pigment that others are, which will probably lead to the production of concentrates of much higher biological activity."

31 J. L. Smith to G. W. Merck, Letter dated October 15, 1941. U.S. National Archives, Record Group No. 227; and Pfizer historical records, compiled by G. B. Stone, vol. III, for the period 1940–1945. "Before we commit ourselves to actively collaborate in this work we must determine through experimentation within our own laboratory what effect the introduction of Penicillium (the one used for preparing Penicillin) has on other fermentation work in which we are already engaged. It has been our experience that the invasion of other mold organisms into our established fermentation work may result in dire consequences. Therefore, we must be sure that the hazards involved are controllable. Our position in this respect has been made clear to all who approached us. During the past few weeks our laboratories have been doing work with Penicillium in a modest way and a considerable number [actually 55] of samples of liquor have been sent to Dr. Dawson for evaluation. Our work so far indicates that many essential factors are still unknown, such as the best composition of the media, the conditions under which Penicillin is produced in satisfactory yield, and the instability of the product after it has been produced in the liquors. The absence of information on these points leads me to believe that research should be undertaken to establish these essential factors before large scale production is attempted."

32 Annual report of the Works Committee to the Board of Directors, dated March 7, 1945; and Pfizer historical records, compiled by G. B. Stone, vol. III, for the period 1940–1945.

33 Pfizer historical records, compiled by G. B. Stone, vol. III, for the period 1940–1945. Report on Brooklyn Activities, 1941, p. 12.

34 J. H. Kane to G. L. Hobby, Letters dated October 2, 3, and 4, 1941; in personal files and laboratory notebooks of G. L. Hobby.

35 Annual report for 1941 from Pfizer historical records, compiled by G. B. Stone, vol. III, for the period 1940–1945.

36 R. D. Coghill, "The Background of Penicillin Production," presented at the Annual Meeting of the American Chemical Society, Cleveland, Ohio, April 5, 1944. Published under the title, "Penicillin—Science's Cinderella," *Chemical & Engineering News* 22 (1944): 588–93.

37 A. N. Richards to K. S. Pilcher, Letter dated April 2, 1943, expressing doubt about use of deep tank fermentation for penicillin production. U.S. National Archives, Records Group No. 227. "The Merck and Squibb people...feel convinced that the deep culture method will displace the individual, shallow culture process. I am not sure they are right. The Chester County Mushroom Laboratories think they could produce 40–60 million units a day if they had a staff working 24 hours a day and they are growing it in small shallow cultures."

38 J. W. Foster and F. E. McDaniel, "Process for the Production of Penicillin." Patent application no. 2,448,790, filed May 15, 1943, issued September 7, 1948. "We have now discovered a method whereby penicillin is produced by penicillin-producing strains of *Penicillium* under submerged conditions. The submerged process of our invention possesses many advantages over the shallow process."

39 R. D. Coghill and A. J. Moyer, "Method for Production of Increased Yields of Penicillin." Patent application no. 2,423,873, filed June 17, 1944, issued July 15, 1947. Assigned to United States of America as represented by the Department of Agriculture. Encompasses the use of phenylacetic acid and salts and esters of phenylacetic acid.

40 A. J. Moyer, "Methods for Production of Penicillin." Patent application no. 2,442,141 and no. 2,443,989, both filed May 11, 1945, issued May 25, 1945. Assigned to United States of America as represented by the Department of Agriculture. Encompasses the use of corn steeps liquor and both surface and submerged growth of the mold.

41 A. J. Moyer, "Method for Production of Penicillin." Patent application no. 2,476,107, filed May 11, 1945, issued July 12, 1949. Assigned to the United States of America as represented by the Department of Agriculture. Encompasses use of lactose–corn steep liquor medium.

42 "Report on Penicillin for the Year 1943," Pfizer Historical Records, compiled by G. B. Stone, vol. III, for the period 1940–1945.

43 "Report on Staff Meeting on Penicillin, held May 10, 1943." Pfizer Historical Records, compiled by G. B. Stone, vol. III, for the period 1940–1945.

44 J. H. Kane to J. L. Smith, Memo dated August 3, 1943, recommending plans be made for conversion to 1,000-gallon tanks; in Pfizer Historical Records, compiled by G. B. Stone, vol. III, for the period 1940–1945.

45 J. E. McKeen to Dayle McClain, Memo dated August 9, 1943, recommending layouts be prepared for tanks of 1,000- to 2,000-gallon capacity; in Pfizer Historical Records, compiled by G. B. Stone, vol. III, for the period 1940–1945.

46 H. W. Florey to J. F. Fulton, Letter dated August 1944. Historical Library, Yale University Medical Library, New Haven, Conn.

47 Annual Report of the Pfizer Works Committee to the Board of Directors, dated March 7, 1945, and summarizing Pfizer activities on penicillin since 1941; in Pfizer Historical Records, vol. III.

48 "Pfizer Expands Penicillin Manufacture—Purchase of Rubel Ice Plant," report dated September 21, 1943; in Pfizer Historical Records, compiled by G. B. Stone, vol. III, for the period 1940–1945.

49 Federal Trade Commission Report, *Economic Report on Antibiotics Manufacture* (Washington, D.C.: U.S. Government Printing Office, June 1958), p. 37.

50 R. D. Coghill, Monthly Reports on Progress toward the Production of Penicillin, prepared for the (U.S.) Medical Research Council, 1941 to 1945. U.S. National Archives, CMR, Record Group No. 227.

51 A. N. Richards to V. Bush, Report on penicillin, dated August 14, 1943. U.S. National Archives, Record Group No. 227.

52 "Food and Drug Requests Samples to Test Every Batch of Penicillin," Food, Drug and Cosmetic Reports, September 4, 1943, p. 4. U.S. National Archives, Record Group No. 227.

53 "Pfizer Co. Penicillin Output at New Peak: Producing 70 Billion Units Monthly," *Journal of Commerce*, June 1, 1944.

54 J. E. McKeen, "The Production of Penicillin," *Trans. Amer. Inst. of Chem. Eng.* 40, no. 6 (December 25, 1944): 747–58.

55 A. L. Elder, ed., "The History of Penicillin Production," *Chem. Eng. Progress Symposium Series* 66, no. 100 (1970): 1–97. Published by the American Institute of Chemical Engineers.

56 The total number of Pfizer employees in 1943 and 1944 was only 919 and 1,134, respectively. Of these, approximately 250 were serving in the armed forces. Thus, almost one-third of available personnel were concentrating their efforts on penicillin production.

57 W. McDermott to G. L. Hobby, Personal communication, 1981, and Hobby, personal recollections.

58 "Cure of Baby Patricia Spurs New Production of Penicillin," *New York Journal-American*, September 12, 1943.

59 "Commercial Solvents Building Large Plant to Produce Penicillin," *Wall Street Journal*, August 20, 1943.

60 "Winthrop Firm to Lift Output of Penicillin," *New York Herald-Tribune*, July 28, 1943.

61 "Winthrop Increases Production of Penicillin," *Chemical Industries*, August 1943.

62 "Report on Squibb-Merck-Pfizer Collaboration, 1944," Pfizer Historical Records, compiled by G. B. Stone, vol. III, for the period 1940–1945.

63 E. P. Abraham and E. B. Chain, "Penicillinase," *Nature, Lond* 146 (1940): 837.

64 "Merck," *Fortune*, June 1947, pp. 217–22.

65 Federal Trade Commission Report, *Economic Report on Antibiotics Manufacture* (Washington, D.C.: U.S. Government Printing Office, June 1958), p. 6.

66 M. Robinson, ed., *A Quiet Man from West Chester*, introduction by Robert D. Coghill (West Chester, Pa.: Chester County Historical Society).

67 In 1945, the contract and the equipment purchased for the penicillin project were sold to the American Home Products Corporation, and in 1946, the building and property of the Chester County Mushroom Laboratories were sold to them also, to be used by Wyeth for control and development facilities.

300

Wyeth Laboratories served a useful function during the 1943 to 1945 period. Not until 1945 did they attempt conversion to submerged fermentation. But at that time, Wyeth installed two experimental tanks for submerged fermentation and gradually made a successful transition to deep tank fermentation. In making the conversion, Wyeth continued to have top priority on delivery of equipment, a priority exceeded only by the Manhattan Project, which they later learned had been the atomic bomb.

68 F. J. Stock, Memorandum on "Office of Civilian Penicillin Distribution," dated April 27, 1944, and sent to all penicillin producers and distributors. U.S. National Archives, Record Group No. 227. "In accordance with the recommendations of the Penicillin Producers Industry Advisory Committee, the W.P.B. [War Production Board] has established an Office of Civilian Penicillin Distribution.... The Office of Civilian Penicillin Distribution will have as its Director Dr. John N. McDonnell who is Chief of the Research Unit, Drugs Branch, W.P.B....Dr. Chester Keefer...during the initial period [will act] ...as Medical Advisor."

69 J. N. McDonnell, War Production Board, Office of Civilian Distribution, Communication H, Memorandum designating depot hospitals for distribution of penicillin for civilian use, dated April 28, 1944. U.S. National Archives, Record Group No. 227.

70 F. J. Stock, War Production Board Report, presented to the American Drug Manufacturers Association, Hot Springs, Va., May 2, 1944. U.S. National Archives, Record Group No. 227. Describes plan for "limited distribution" of penicillin to a selected list of hospitals.

71 F. J. Stock, "Status of Drug Supplies," presented to the American Pharmaceutical Manufacturers Association, Waldorf-Astoria Hotel, New York City, December 12, 1944. U.S. National Archives, Record Group No. 227. By December 1944, 2,700 hospitals were receiving monthly quotas of penicillin and were regularly placing orders for their needs. "A year ago, the average potency of penicillin was 100 to 200 units per milligram. Today every lot exceeds 350 units per milligram and a high percentage has a potency of 800 to 1100 units per milligram.... During May [1944], the first month of civilian distribution, a total of 12.2 billion units...were released. This quantity ...[has] increased every month...up to 19.3 billion units in August... 26.8 billion units in November...up to approximately 30 billion units [in December].... Civilians are now receiving penicillin valued at $750,000 per month."

Chapter 10

1 L. Ettlinger, "Wartime Research on Penicillin in Switzerland and Antibiotic Screening," in *The History of Antibiotics—A Symposium*, ed. J. Parascandola (Madison, Wis.: American Institute of the History of Pharmacy, 1980), pp. 57–68.

2 G. Nesemann, "Mikrobiologie in Hoechst," translation of address given at the

dedication of new laboratories and pilot plant of the Microbiology Division of Hoechst A/G, January 31, 1975.

3 B. Elema, *One Hundred Years of Yeast Research* (Delft: Koninklijke Nederlandsche Gist- en Spiritusfabriek N.V., 1970), pp. 35–49.

4 P. Mars (Gist-Brocades N.V.), Personal communication to Dr. F. X. Murphy, Pfizer, Inc., dated January 3, 1979.

5 W. Th. Nauta, H. K. Oosterhuis, A. C. Van der Linden, P. Van Duin, and J. W. Dienske, "The Structure of Expansine, a Metabolic Product of Penicillium Expansum Westl. with Antibiotic Properties," *Recueil des Travaux Chimiques des Pays-Bas* (Amsterdam), T. 64, 3/4 (September–October 1945): 254–55.

6 W. Th. Nauta, H. K. Oosterhuis, A. C. Van der Linden, P. Van Duin, and J. W. Dienske, "On the Structure of Expansine: A Bactericidal and Fungicidal Substance in *Penicillium expansum Westl.*," *Recueil des Travaux Chimiques des Pays-Bas* (Amsterdam: La Soc. Chim. Neerl.), T. 65, 12 (December 1946): 865–76.

7 "The World of Fermentation," *Quarterly International Journal of Gist-Brocades N.V.*, no. 3 (November 1978): 3–5.

8 Mr. Koetschet (assistant scientific director, Rhône-Poulenc-Santé), "Penicillin of Société Rhône-Poulenc," translation of lecture delivered at general internal meeting, May 5, 1947. Received by Dr. G. L. Hobby, courtesy of Mr. L. Girardet of Rhône-Poulenc-Santé on December 12, 1978.

9 G. H. Werner (Rhône-Poulenc, Centre Nicolas Grillet), Personal communication to Dr. G. L. Hobby, dated November 28, 1978.

10 "Penicillin Manufacture in France," *Chemist and Druggist*, December 8, 1945, p. 615.

11 H. Kuntscher, "The History of Biochemie Ges. m.b.H. as Reflected in Its Research and Production," unpublished manuscript, 1972.

12 E. Brandl, "Biochemie GmbH im Spiegel ihrer Forschung und Entwicklung," *Subsidia medica* 27, nos. 1–2 (1975): 9–11.

13 Among those who visited Germany at the close of World War II were Dr. E. C. Kleiderer of Eli Lilly & Co., (the Lilly Research Laboratories), Dr. J. B. Rice of Winthrop Chemical Corporation, Dr. J. H. Williams of the American Cyanamid Corporation, and Dr. Victor Conquest of Armour Laboratories in Chicago.

14 "Germans Lagged on Penicillin, U.S. Report on I. G. Reveals," *Journal of Commerce and Commercial*, New York: Friday, September 21, 1945.

15 Manfred Kiese, "Chemotherapie mit Antibakteriellen Stoffen aus Niederen Pilzen und Bakterien," *Klinische Wochenschrift*, 22, nos. 32–33 (August 7, 1943): 505–11.

16 Y. Yagisawa, "Early History of Antibiotics in Japan," in *The History of Antibiotics—A Symposium*, ed. J. Parascandola (Madison, Wis.: American Institute of the History of Pharmacy, 1980), pp. 69–90.

17 R. D. Defries to A. N. Richards, Letter dated October 1, 1943. U.S. National Archives, Record Group No. 227.

18 P. H. Greey and C. H. Best, Extract from Report on Penicillin to Sub-

committee on Infection, Associate Committee on Medical Research, National Research Council, Ottawa. Project Med. 20, dated July 16, 1943. U.S. National Archives, Record Group No. 227.

19 R. D. Coghill and R. S. Koch, "Penicillin: A Wartime Accomplishment," *Chemical & Engineering News* 23, no. 24 (December 25, 1945): 2310–16.

Chapter 11

1 Albert Elder, "Penicillin," *Scientific Monthly* 58 (June 1944): 405–09.
2 "Penicillin Data Exchange Exempted from Antitrust Laws," *Oil, Paint and Drug Reporter*, December 20, 1943.
3 J. P. Baxter, 3d, "Penicillin," in *Scientists against Time* (Boston: Little, Brown & Co., 1947), ch. 22.
4 "Knighthoods for Penicillin Pioneers," *Chemical Trade Journal and Chemical Engineer*, June 9, 1944. The knighthoods were awarded to Fleming as the "discover of penicillin" and to Florey "for services in development of penicillin."
5 G. Liljestrand, "Physiology of Medicine 1945," Presentation Speech of Professor G. Liljestrand, Member of the Staff of Professors of the Royal Carolina Institute, in *Nobel Lectures, Including Presentation Speeches and Laureates' Biographies, Physiology or Medicine, 1943–1962* (Amsterdam, London, and New York: Elsevier Publishing Company, 1964).
6 J. W. Fulton, Letter to P. J. Hedenius (Swedish Legation, New York) dated October 7, 1944; cable to G. Liljestrand, dated October 25, 1944; letter to G. Liljestrand dated November 8, 1944. "Our recommendation," Fulton wrote, "would be [to award it to], in order to preference: [1] Florey alone, [2] Florey and Chain, [3] Florey, Chain, and Fleming." This expression of opinion, according to Fulton, represented the "considered opinion of a large group in this country [the U.S.] including the Division of Medical Sciences of the National Research Council and the officers of the American Medical Association." The enthusiasm with which Fleming was received in the United States a few months later in 1945 and the credit given him then for his discovery of penicillin makes one wonder how large a group in the United States was actually in agreement with Fulton on this matter. Fortunately, Liljestrand and his committee were influenced only by their personal knowledge and by available published information.
7 A. Fleming, "Penicillin," Nobel Lecture, December 11, 1945, in *Nobel Lectures*.
8 H. W. Florey, "Penicillin,' Nobel Lecture, December 11, 1945, in *Nobel Lectures*.
9 E. B. Chain, "The Chemical Structure of the Penicillins," Nobel Lecture, March 20, 1946, in *Nobel Lectures*.
10 E. P. Abraham, W. Baker, E. Chain, H. W. Florey, E. R. Holiday, and R. Robinson, "Nitrogenous Character of Penicillin," *Nature, Lond.* 149 (1942): 356.

11 H. T. Clarke, J. R. Johnson, and Sir Robert Robinson, eds., "Brief History of the Chemical Study of Penicillin," in *The Chemistry of the Penicillins* (Princeton, N.J.: Princeton University Press, 1949), pp. 3–9. In February 1944, agreement for exchange of information between the British and American chemists was reached. In Great Britain, the Medical Research Council formed the Penicillin Synthesis Committee. In America, the Office of Scientific Research and Development delegated Dr. Hans T. Clarke of Columbia University to coordinate the chemical research on penicillin in the United States. These two bodies agreed to exchange reports at monthly intervals, beginning April 1944. This exchange program continued until October 1945. More than seven hundred reports were circulated during that time and provided the scientific data that were reported in this monograph.

12 H. B. MacPhillamy, O. Wintersteiner, and J. F. Alicino, "A Crystalline Sodium Penicillin," Squibb Report of July 27, 1943, to Dr. H. T. Clarke's Chemistry of Penicillin Committee, The O.S.R.D.

13 The Editorial Board of the Monograph on the Chemistry of Penicillin: "Biosynthesis of Penicillins," *Science* 106, no. 2761 (November 28, 1947): 503–05.

14 O. K. Behrens, J. Corse, R. G. Jones, M. J. Mann, G. F. Soper, F. R. Van Abeele, and M. Chiang, "Biosynthesis of Penicillins. Biological Precursors for Benzyl Penicillin (Penicillin G)," *J. Biol. Chem.* 175 (1948): 751. See also O. K. Behrens, "Biosynthesis of Penicillins," in *The Chemistry of Penicillin*, ed. H. T. Clarke, J. R. Johnson, and R. Robinson (Princeton, N.J.: Oxford University Press, 1949), p. 657.

15 D. Crowfoot, C. W. Bunn, B. W. Rogers-Low, and A. Turner-Jones, "The X-Ray Crystallographic Investigation of the Structure of Penicillin," in *The Chemistry of Penicillin*, ed. Clarke, Johnson, and Robinson, p. 310.

16 G. T. Stewart, *The Penicillin Group of Drugs* (Amsterdam, London, and New York: Elsevier Publishing Co., 1965), pp. 17–18.

17 K. Kato, *J. Antibiotics* (Tokyo), A6, 130 (1953): 184–.

18 J. C. Sheehan, *The Enchanted Ring—The Untold Story of Penicillin* (Cambridge, Mass., and London: MIT Press, 1982).

19 J. C. Sheehan, "Bicyclic Lactams." Patent application no. 2,721,196, filed January 5, 1953, issued October 18, 1955. A continuation in part...of prior copending application, serial no. 176,013, filed July 26, 1950, now abandoned. Assigned to Bristol Laboratories. "This invention...constitutes the first chemical evidence from the synthetic side, for the beta-lactam formulation of penicillin.... This invention constitutes a claim for a 'process for the preparation of bicyclic beta-lactams.'"

20 F. R. Batchelor, E. B. Chain, and G. N. Rolinson, "6-Aminopenicillanic Acid. I. 6-Aminopenicillanic Acid in Penicillin Fermentations," *Proc. Royal Society*, B, 154 (1961): 478–89.

21 F. P. Doyle, J. H. C. Nayler, and G. N. Rolinson, "Recovery of Solid 6-Aminopenicillanic Acid," patent application no. 2,941,995, filed July 22, 1958, issued June 21, 1960. Assigned to Beecham Research Laboratories. "The present invention...includes a process for the preparation of 6-amino-penicillanic acid or its salts in which a penicillin-producing mold is grown in a

nutrient medium and the 6-aminopenicillanic acid or a salt thereof is isolated from the fermentation liquor obtained." Claims priority in Great Britain filed August 2, 1957.

22   H. G. Lazell, *From Pills to Penicillin—the Beecham Story* (London: Heinemann, 1975).

23   G. N. Rolinson, F. R. Batchelor, D. Butterworth, J. Cameron-Wood, M. Cole, G. C. Eustace, M. V. Mart, M. Richards, and E. B. Chain, "Formation of 6-Aminopenicillanic Acid with Penicillin by Enzymatic Hydrolysis," *Nature* 187 (1960): 236–37.

24   G. N. Rolinson, "6-APA and the Development of the Beta-Lactam Antibiotics," *J. Antimicrobial Chemotherapy* 5 (1979): 7–14.

25   E. Menotti to G. L. Hobby, Letter dated November 3, 1978, summarizing the role of Bristol Laboratories in the development of the penicillins.

26   L. C. Cheney, "Beta-Lactam Antibiotics—Some Recent Developments," presented at the meeting of the American Chemical Society, Division of Medicinal Chemistry, Atlantic City, N.J., September 8–13, 1968. "In August 1943, Bristol was one of the plants selected by the War Production Board for wartime penicillin production. Bristol-Myers had just purchased the Cheplin Biological Laboratories, a small acidophilus milk and generic pharmaceutical producer in Syracuse. I [Emil Menotti] had just arrived to organize Research and Control so a technical staff of one was hardly a major force for the construction of a new plant. Strangely enough though, Bristol is still a major producer while the majority of the other 21 authorized plants...have since gone out of existence. We were authorized to build a plant to handle 80,000 sq. 2 qt. milk bottles to be on a continuous conveyor going thru a fermentation-extraction cycle of 14–18 days. Just before start-up, the feasibility of tank fermentation...reared its head so the bottle train was ripped out and twenty-four 2,400-gallon tanks [were] crammed into the same quarters. An interesting sidelight on Penicillin V at this time: During our laboratory production of surface culture penicillins F, DHF [dihydro-F], etc. we changed strains and suddenly a portion of the activity in our broth was no longer extractable in chloroform. One of our chemists—of which we had three—asked me what we should call the chloroform unextractable activity. We were able to extract it with less lyopholic solvents and lyophilize the material. I suggested we call it 'Substance X' since we were not certain it was a penicillin. We accumulated a substantial quantity and sent it to Robert Coghill at Peoria NRRL where the group crystallized P-hydroxy G from this 'Substance X' labelled residuals and for want of a better name called it 'Penicillin X.' Of course, Bristol went through the crisis in fermentation later when a change to...[a strain of *P. chrysogenum*] was found to produce large quantities of Penicillin K. At one Production Penicillin Meeting, it was disclosed that phenylacetic acid added to the fermentation gave the familiar [penicillin] G. In 1959, Bristol and Beecham entered into an agreement relating to 6-APA whence followed the string of semi-synthetic penicillins on the market. Bristol also sponsored John Sheehan's work at MIT from 1948 to approximately 1960 resulting in the first totally synthetic penicillin and also synthesis of 6-APA."

27 L. C. Cheney, "Antibiotics," in *Organic Chemistry, an Advanced Treatise*, 4 vols. (New York: John Wiley & Sons, 1953), 3: 533–52.
28 L. C. Cheney, "Some Recent Advances in Penicillin-Cephalosporin Chemistry," presented at the Department of Medicinal Chemistry, State University of New York at Buffalo, November 1, 1965.
29 T. R. E. Hoover and G. L. Dunn, "The Beta-Lactam Antibiotics," in *Burger's Medicinal Chemistry*, 4th ed.

Chapter 12

1 C. S. Keefer and W. L. Hewitt, *The Therapeutic Value of Streptomycin* (Ann Arbor, Mich.: J. W. Edwards, 1948). See also *J. Amer. Med. Assn.* 132 (September 7, 1946): 4–11; 132 (September 14, 1946): 70–77. The streptomycin industry producers at the time included Abbott Laboratories; Commercial Solvents Corp.; Heyden Chemical Corp.; Eli Lilly & Co.; Merck & Co.; Parke, Davis & Co.; Chas. Pfizer & Co.; Schenley Laboratories; E. R. Squibb and Sons; The Upjohn Co.; and Wyeth.
2 V. Bush, "Science—The Endless Frontier," Report to the President (July 1945) (Washington: U.S. Government Printing Office, 1945).
3 W. McDermott with D. E. Rogers, "Social Ramifications of Control of Microbial Disease," *Johns Hopkins Medical Journal* 151 (1982): 302–12.
4 W. McDermott, "Pharmaceuticals: Their Role in Developing Societies," *Science* 209 (July 11, 1980): 240–45.
5 R. Austrian, "Pneumococcal Infection and Pneumococcal Vaccine," *New England Journal of Medicine* 297, no. 17 (October 27, 1977): 938–39.
6 J. D. Williams and A. Howard, eds., and G. G. Grassi, guest ed., "Bacterial Resistance." Proceedings of an international meeting held at Taormina, Sicily, October 14, 1976. Published in *The Journal of Antimicrobial Chemotherapy*. Supplement C to vol. 3, November 1977.
7 M. Cole, "Inhibitors of Bacterial Beta-Lactamases," *Drugs of the Future* 6, no. 11 (1981): 697–727.
8 K. Bush and R. B. Sykes, "Beta-Lactamase Inhibitors in Perspective," *J. Antimicrob. Chemother.* 11 (1983): 97–107.
9 J. L. Maddocks, "Indirect Pathogenicity," *J. Antimicrob. Chemother.* 6 (1980): 307–09.
10 L. K. Altman, "The Doctors' World: Infections Still a Big Threat," *New York Times*, July 20, 1981, sec. C, p. 2
11 A. Demain and H. A. Solomon, "Industrial Microbiology," *Scientific American* 245, no. 5 (September 1981): 67–76.
12 A. Demain, "Industrial Microbiology," *Science* 214 (January 27, 1981): 987–95.
13 A. L. Demain, "New Applications of Microbial Products," *Science* 219, no. 4585 (February 11, 1983): 709–14.
14 Y. Aharonowitz and G. Cohen, "Microbiological Production of Pharmaceuticals," *Scientific American* 245, no. 3 (September 1981): 140–52.

15 D. E. Eveleigh, "Microbiological Production of Industrial Chemicals," *Scientific American* 245, no. 3 (September 1981): 154.

16 J. H. Vorernakes and R. P. Elander, "Genetic Manipulation of Antibiotic-Producing Microorganisms," *Science* 219, no. 4585 (February 11, 1983): 303–09.

17 R. Sherwood and T. Atkinson, "Genetic Manipulation in Biotechnology," *Chemistry & Industry*, April 4, 1981, pp. 241–47.

18 F. Adams, Personal communication, 1983.

19 "ASM Files Amicus Brief on Patenting Microorganisms," *ASM News* 46, no. 3 (1980): 108–11.

20 R. H. Shryock, *American Medical Research, Past and Present* (New York: Commonwealth Fund, 1947), pp. 140–45.

21 I. P. Cooper, *Biotechnology and the Law* (New York: Clark Boardman Co., 1982). The wisdom of attempting to patent penicillin was a subject of concern even in 1940 and 1941. At the time it was generally agreed that—since penicillin is a natural product, produced spontaneously in nature—a patent on the substance would be out of the question and certainly unethical. The feasibility of obtaining patent protection on a microbial strain that produced penicillin seemed remote. But, by 1944, thinking had changed and streptomycin, also a natural product, was patented by Waksman and his associates. Today, the invention of novel microorganisms by recombinant DNA processes, the genetic manipulation of microorganisms to yield strains with special properties, and patent protection of new strains are commonplace. See S. A. Waksman and A. Schatz, *Streptomycin and Process of Preparation*, patent application no. 2,449,866, filed February 9, 1945, issued September 21, 1948. Assigned to the Rutgers Research and Endowment Foundation, a nonprofit corporation of New Jersey. See also G. Sermonti, "Mutation and Mutation Breeding," in *Genetics of Industrial Microorganisms*, ed. O. K. Sebek and A. I. Larkin (Washington, D.C.: American Society for Microbiology, 1979).

22 M. Lee, "The Economics of Research-Based Industry," in *Science, Industry and the State*, ed. G. Teeling-Smith (Oxford, London, and New York: Symposium Publications Division, Pergamon Press, 1965).

Sources

Major Sources

American Institute of the History of Pharmacy, University of Wisconsin, Madison, Wisconsin.

The Fleming Papers, British Library, London, England.

The Florey Archives, Library of the Royal Society, London, England.

The National Archives, United States of America, Washington, D.C., Record Group No. 227, OSRD (CMR).

General References

Anderson, D. G., and Keefer, C. S. *The Therapeutic Value of Penicillin: A Study of 10,000 Cases*. Ann Arbor, Mich.: J. W. Edwards, 1948.

Bickel, L. *Rise up to Life: A Biography of Howard Walter Florey Who Gave Penicillin to the World*. London: Angus and Robertson, 1972.

Clarke, H. T., Johnson, J. R., and Robinson, Sir Robert, eds. *The Chemistry of Penicillin*. Princeton, N.J.: Princeton University Press, 1949.

Dowling, H. F. *Fighting Infection: Conquests of the Twentieth Century*. A Commonwealth Fund Book. Cambridge, Mass., and London: Harvard University Press, 1977.

Dubos, R. *The Professor, The Institute, and DNA*. New York: Rockefeller University Press, 1976.

Epstein, S., and Williams, B. *Miracles from Microbes: The Road to Streptomycin*. New Brunswick: Rutgers University Press, 1946.

Federal Trade Commission, Economic Report on Antibiotics Manufacture. Washington, D.C.: U.S. Government Printing Office, June 1958.

Fleming, A., ed. *Penicillin and Its Practical Application.* 2d ed. London: Butterworth & Co.; St. Louis: C. V. Mosby Co., 1950.

Florey, H. W., Chain, E. B., Heatley, N. G., et al. *Antibiotics—A Survey of Penicillin, Streptomycin, and Other Antimicrobial Substances from Fungi, Actinomycetes, Bacteria, and Plants.* 2 vols. London: Oxford University Press, 1949.

Florey, M. E. *The Clinical Application of Antibiotics (Penicillin).* London: Oxford University Press, 1952.

Garrod, L. P., and O'Grady, F. *Antibiotic Chemotherapy.* 3d ed. Baltimore: Williams and Wilkins Co., 1972.

Hare, R. *The Birth of Penicillin, and the Disarming of Microbes.* London: George Allen and Unwin, 1978.

————. "The Scientific Activities of Alexander Fleming, other than the Discovery of Penicillin." *Medical History* 27 (1983): 347–72.

Herrell, W. *Penicillin and Other Antibiotic Agents.* Philadelphia and London: W. B. Saunders Co., 1945.

Himmelweit, F., ed., with the assistance of Martha Marquardt, under the editorial direction of Sir Henry Dale. *The Collected Papers of Paul Ehrlich.* 3 vols. London: Pergamon Press, 1956, 1957, 1960.

Lazell, H. G. *From Pills to Penicillin.* London: William Heinemann, 1975.

Long, P. H., and Bliss, E. A. *The Clinical and Experimental Use of Sulfanilamide, Sulfapyridine, and Allied Compounds.* New York: MacMillan Co., 1939.

Ludovici, L. J. *Fleming—Discoverer of Penicillin.* London: Andrew Dakers, 1952.

Macfarlane, G. *Howard Florey: The Making of a Great Scientist.* Oxford: Oxford University Press, 1979.

————. *Alexander Fleming—The Man and the Myth.* Cambridge, Mass.: Harvard University Press, 1984.

Matthews, L. G. *History of Pharmacy in Britain.* Edinburgh and London: E. & S. Livingstone, 1962.

Maurois, A. *The Life of Sir Alexander Fleming.* New York: E. P. Dutton & Co., 1959.

Middleton, W. S. *Values in Modern Medicine.* Madison: University of Wisconsin Press, 1972.

Moore, J. E. *Penicillin in Syphilis.* Springfield, Ill.: Charles C. Thomas, 1946.

Noble, W. C. *Coli: Great Healer of Men. The Biography of Dr. Leonard Colebrook, FRS.* London: William Heinemann Medical Books, 1974.

Parascandola, J., ed. *The History of Antibiotics.* A symposium sponsored by the Divisions of History of Chemistry and Medicinal Chemistry, American Chemical Society, April 5, 1979. Madison, Wis.: American

Institute of the History of Pharmacy, 1980.

Ratcliff, J. D. *Yellow Magic—The Story of Penicillin*. New York: Random House, 1945.

Ross, W. S. *The Life/Death Ratio: Benefits and Risks in Modern Medicines*. New York: Reader's Digest Press, 1977.

Schwartzman, D. *Innovation in the Pharmaceutical Industry*. Baltimore: Johns Hopkins University Press, 1976.

Sheehan, J. C. *The Enchanted Ring: The Untold Story of Penicillin*. Boston: MIT Press, 1982.

Shryock, R. H. *American Medical Research, Past and Present*. New York: Commonwealth Fund, 1947.

Sokoloff, B. *The Story of Penicilin*. Chicago and New York: Ziff-Davis Publishing Co., 1945.

Stewart, G. T. *The Penicillin Group of Drugs*. Amsterdam, London, and New York: Elsevier Publishing Co., 1965.

Teeling-Smith, G., ed. *Science, Industry, and the State*. London: Symposium Publications Division, Pergamon Press, 1965.

———. *Economics and Innovation in the Pharmaceutical Industry*. London: Office of Health Economics, 1969.

Thomson, A. Landsborough. *Half a Century of Medical Research*. Vol. I, *Origins and Policy of the Medical Research Council (UK)*; Vol. II, *Programme of the Medical Research Council (UK)*. London: H.M.S.O., 1973, 1975.

Waksman, S. A. *Microbial Antagonisms and Antibiotic Substances*. New York: Commonwealth Fund, 1945.

———. *My Life with the Microbes*. New York: Simon and Schuster, 1954.

———. *The Antibiotic Era*. Tokyo: Waksman Foundation of Japan, 1975.

Welch, H. and Lewis, C. N. *Antibiotic Therapy*. Washington, D.C.: Arundel Press, 1951.

Wilson, D. *In Search of Penicillin*. New York: Alfred A. Knopf, 1976.

Work, T. S. and Work, E. *The Basis of Chemotherapy*. New York: Interscience Publishers, 1948.

Index

313

Index

Barnard Castle, Glaxo penicillin plant at, 136
Beaverbrook, Lord William, 214, 215
Beecham Laboratories: new, tailor-made penicillins from, 228
Behrens, Dr. Otto, 221, 224, 230
Beijerinck, Dr. N. W., 202
Benzyl penicillin (penicillin G): crystalline sodium salt, 218; para-gydroxyphenyl acetic acid as precursor for, 221
Beta-lactam antibiotics, penicillin type and cephalosporin type, 221
Bioconversion of antibiotics, 242
Biosynthesis of penicillins, 217, 221, 224, 228, 230, 231; penicillinase-resistant penicillins with gram negative activity, 217, 230, 231
Bogert, Dr. V., 183
Boon, Dr. W. R., 135–38
Boots Pure Drug Company, 128–30
Bornstein, Dr. S., 37
Boulogne-sur-Mer, Fleming laboratories at in W. W. I, 24, 36
Boyd, Col. J. S. K., 122
Bradley Technological Institute, 92
Bristol Laboratories, 305n26
British Drug Houses, Ltd., 128–30
British Industries Act of *1948,* 130
British Ministry of Health, 25
Burroughs Wellcome Company, 128–30
Bush, Dr. Vannevar, 92, 104, 105, 141; outlines basis for establishment of National Science Foundation, 233–34
Bushnell General (Army) Hospital, 149

Cairns, Brig. Gen'l. Hugh: clinical evaluation of penicillin in Algiers, 124
Cambridge: Sir William Dunn School of Bio-chemistry at, 49
Canada: penicillin production in, 210
Carbenicillin, 231
Carter, Dr. Howard, 177
Caution, in the industry, 298n21
Cell walls of bacteria, destruction of by lysozyme, 15
Cephalosporin, 90; production by Cephalosporium acremonium Streptomyces Spp., 231
Cephamycin, 231
Chain, Prof. E. B.: microbial antagonisms, literature survey, rediscovery of penicillin, 4, 19, 48; studies on lysozyme, 16, 18; mucolytic enzymes, 31; hyaluronidase, 69; application to Rockefeller Foundation, support for study of mucinases, 22

—Penicillin research, 47, 59, 78, 106, 125; first report on, 47, 58, 65, 125; second report on, 66; biosynthetic, 228; Nobel prize recipient for, 215
Chemical studies, coordination of with MRC, 287n36
Chemotherapy: effectiveness and potential of, 23, 33, 34; basic principles established, 28–33; puerperal sepsis, treatment of, 33
Chesney, Dr. Alan, 159
Chester County Mushroom Laboratories, 135, 186, 191–95, 300nn66, 67
Christie, Dr. R.: clinical trials of penicillin in Great Britain, 116
Clark, Dr. W. Mansfield, 103, 104
Clarke, Dr. Hans T., 94, 304nn11, 14
Clinical evaluation of penicillin, organization for in W. W. II, 92–93
Clostridium septique, 66
Cloxacillin, 231
Clutterbuck, Dr. P. W., 44, 47, 87, 106
Coburn, Dr. A., 174
Coghill, Dr. R. D., 88, 94–103, 135, 185
Colebrook, Dr. Leonard: development of sulfonamides, 24, 25; treatment of puerperal sepsis, 33
Columbia University College of Physicians and Surgeons, 16; studies on penicillin, 71–80, 103, 173–78, 273n2, 295n12
Commercial Solvents Corporation: submerged growth fermentation plant in U.S., penicillin production, 137
Committee on Medical Research (CMR), 92, 94; agreement with penicillin producers (*1942*), use of single agency for evaluation, 94
Connective tissue, ground substance: hyaluronic acid in, 21; hyaluronidase action on, 21
Craddock, Dr. S.: studies on penicillin with A. Fleming, 37–44
Cragwall, Gordon, 73, 74, 176
Crystallographic studies on penicillin, 221

D-Day: penicillin supplies for, 85, 131, 145, 213
Dale, Sir Henry, 128–30, 287n34
Dawson, Dr. M. H.: hyaluronic acid in streptococci, 21; transformation of pneumococcal types, 26–28; penicillin, first purposeful study, 69–80, 106, 125, 161–70, 173, 175, 176; death of, 77; therapeutic response an immediate goal of, 72; first one hundred treated patients, 289n20, 295n12

314

Index

—Submerged fermentation (*continued*)
tion, *1943 to 1945,* 196; civilian distribution, 195–97; in Great Britain, 282*n*16; general penicillin committee, 287*n*36; penicillin producers conference, 287*n*26; PEN Reports, 287*n*36; penicillin output, 298*n*31; penicillin synthesis committee, 304*n*11

—Therapeutic efficacy, 273*n*2; acute infections, 146–49; surgical and wound infections, 149–51; syphilis, 152–58; subacute bacterial endocarditis, 161–70; absorption and pharmacological action, 293*n*68. *See also* Medical Research Council

Producer strain mutations, 100–01, 234

—Biosynthesis of, 228–31; 284*nn*15, 18; 304*n*26

—First commercial plant, 187–90

—Process patents on, 284*n*18, 299*nn*38–41, 304*nn*19–20, 307*n*21

—Structure and chemical synthesis of, 214–24, 224–28, 287*n*36, 293*n*67, 298*n*30

Pfizer, Inc. (Chas. Pfizer & Co., Inc.): 103, 104, 105, 149, 165–70, 187–90, 300*n*56

Pulvertaft, Lt. Col. R. J. V., 122–24

Raistrick, Dr. Harold, 44, 47, 129, 132
Rake, Dr. Geoffrey, 79
Raper, Dr. K.: biological properties of molds and strain selections, 87, 88, 94–96, 100, 275*n*37
Ratajak, R. E., 183
Regna, R. P., 183
Reid, Dr. Roger, 46
Resistance, induced, to penicillin, 79, 107, 108; natural, to penicillin, 294*n*68
Rettew, Dr. G. R., 191–94
Rheumatoid arthritis, 23
Richards, Prof. A. N., 92, 104, 125, 141, 279*n*19, 299*n*37
Richards, Dr. D., 174
Roberts, Dr. E. A. H., 18, 60
Robinson, Dr. F. A., 128–30
Robinson, Sir Robert, 60, 132
Rockefeller Foundation, Natural Sciences Division: plan to advance experimental pathology, 19, 59, 67, 68
Rockefeller Institute for Medical Research, 28, 35
Rolinson, Dr. G., 228
Royal Air Force Burns Unit: evaluation of penicillin for topical treatment of burns, 121–22

St. Mary's Hospital Medical School (and Inoculation Department), 24, 25

318

Salvarsan, neosalvarsan: synthesized, 24; antisyphilitic action, 30
Sanders, N. G., 60
Sanochrysin (gold compound), 24
Schering Corporation, 135
Sheehan, Dr. J. C.: total chemical synthesis, 224, 225
Shull, Dr. G. M., 183
Sicily, invasion of: penicillin, MRC clinical trials of, 124
Simmons, Brig. Gen'l. J. S., 92
Skin permeability, increase by hyaluronidase, 21
Smith, Rear Admiral H. W., 92
Smith, John L.: Pfizer research, development, production, 72–74, 159, 171–90
Smith, Dr. W. J., 183
Spencer, Steven, 73, 176
Spreading factor: from invasive hemolytic streptococci, 21; intesticular extracts, 21
Squibb, E. R., & Sons, Inc. (Squibb Institute for Medical Research), 79, 104, 134–40, 171, 175, 178–87, 218
Standard, for penicillin, 77, 275*n*37
Staphylococcus aureus, 66
Stearns (Frederick Stearns, Inc.), 135
Steep liquor (corn), penicillin culture media, 89, 90, 96, 99, 101
Stewart, Dr. Irwin, 92
Stock, F., 142–45
Stodola, Dr. Frank, 218
Streptococci, hemolytic (*Streptococcus pyogenes*): serological classification of, 25–28
Substrate of lysozyme, 18
Sulfadiazine, 35
Sulfamethazine, 35
Sulfanilamide, 269*n*57
Sulfapyridine (M & B 693), 34, 129
Sulfathiazole, 35
4'-Sulfonamido-2,4-diaminodiazo benzene (prontosil), 32
Sulfonamides: development of, 24, 25; antimicrobial activity of, 31–35; staphylococcal infections resistant to, 119; chemical nature of, 269*n*58; Fleming studies on, 272*n*107
Sulfur, in penicillin molecule, 218
Supply, Ministry of: authorizes construction of submerged growth penicillin plants, 134–40
Synthesis (chemical): of para aminobenzene sulfonamide, 31; of penicillin V, 225

Testicular extracts: source of spreading factor, 21